HOW TO PASS

HIGH PHYSICS

Hugh McGill

HODDER GIBSON
AN HACHETTE UK COMPANY

Every effort has been made to trace all copyright holders, but if any have been inadvertently overlooked the Publishers will be pleased to make the necessary arrangements at the first opportunity.

Although every effort has been made to ensure that website addresses are correct at time of going to press, Hodder Gibson cannot be held responsible for the content of any website mentioned in this book. It is sometimes possible to find a relocated web page by typing in the address of the home page for a website in the URL window of your browser.

Hachette UK's policy is to use papers that are natural, renewable and recyclable products and made from wood grown in sustainable forests. The logging and manufacturing processes are expected to conform to the environmental regulations of the country of origin.

Orders: please contact Bookpoint Ltd, 130 Milton Park, Abingdon, Oxon OX14 4SB. Telephone: (44) 01235 827720. Fax: (44) 01235 400454. Lines are open 9.00–5.00, Monday to Saturday, with a 24-hour message answering service. Visit our website at www.hoddereducation.co.uk. Hodder Gibson can be contacted direct on: Tel: 0141 848 1609; Fax: 0141 889 6315; email: hoddergibson@hodder.co.uk

© Hugh McGill 2011

First published in 2011 by
Hodder Gibson, an imprint of Hodder Education,
An Hachette UK Company
2a Christie Street
Paisley PA1 1NB

Impression number   5 4 3 2 1
Year   2013 2012 2011

All rights reserved. Apart from any use permitted under UK copyright law, no part of this publication may be reproduced or transmitted in any form or by any means, electronic or mechanical, including photocopying and recording, or held within any information storage and retrieval system, without permission in writing from the publisher or under licence from the Copyright Licensing Agency Limited. Further details of such licences (for reprographic reproduction) may be obtained from the Copyright Licensing Agency Limited, Saffron House, 6–10 Kirby Street, London EC1N 8TS.

Cover photo © Photodisc/Getty Images
Illustrations by GreenGate Publishing Services
Typeset in Gill Sans Alternative by GreenGate Publishing Services, Tonbridge, Kent
Printed in Great Britain by CPI Antony Rowe

A catalogue record for this title is available from the British Library

ISBN: 978 1444 120486

Contents

Mechanics and properties of matter
1. Scalars and vectors
2. Vector components
3. Adding vectors
4. Acceleration
5. Graphs of motion
6. The equations of motion
7. Using the equations of motion
8. Projectiles
9. Balanced and unbalanced forces
10. Force, mass and acceleration
11. Analysing forces
12. Energy and work
13. Momentum
14. Impulse
15. Density
16. Pressure
17. Pressure in fluids
18. Buoyancy force
19. Temperature
20. Kinetic model of matter and gas pressure
21. Gas laws
22. More gas laws and kinetic model of matter

Electricity and electronics
23. Electric fields
24. Charge, work and potential difference
25. E.m.f. and terminal potential difference
26. E.m.f. and internal resistance
27. Energy in electrical circuits
28. Power in electrical circuits
29. Current, voltage and resistance in series circuits
30. Current, voltage and resistance in parallel circuits
31. Current, voltage and resistance in more complex circuits
32. Potential divider
33. Wheatstone bridge
34. Alternating current and voltage
35. Energy and power in resistive circuits
36. Capacitance 1
37. Capacitance 2
38. Capacitors in d.c. circuits
39. Capacitors in a.c. circuits
40. Uses of capacitors
41. Operational amplifiers
42. Op-amps in circuits 1
43. Op-amps in circuits 2

Radiation and matter
44. Wave properties
45. Wave definitions
46. Wave behaviour
47. Interference
48. Investigating interference
49. Gratings
50. Spectra
51. Refraction 1
52. Refraction 2
53. Total internal reflection and critical angle
54. Irradiance
55. Photons
56. Photoelectric effect
57. Energy levels in atoms and line spectra
58. Lasers
59. Semiconductors
60. p–n junction diode
61. Applications of diodes
62. n-channel enhancement MOSFET
63. The atom and nucleus
64. Nuclear reactions
65. Dosimetry
66. Background radiation
67. Radiation and safety

Course skills
68. Numbers and units
69. Uncertainties
70. Random and systematic uncertainties

Layout of answers to numerical questions

Please note that space limitations in this book have meant that we can't follow 'best practice' advice on laying out answers to numerical questions. General advice on this is included in the 'Welcome' section of *How to Pass Higher Physics* by the same author. In particular, see also the solutions to Exercise 8, Questions 1–3 and all of the solutions to numerical questions in that title (ISBN 978 0340 974025).

Higher Physics
Mechanics and properties of matter

Scalars and vectors

Q1 Define the terms *scalar* and *vector*.

Q2 Which of the following quantities are scalars and which are vectors?

acceleration distance force mass
momentum pressure volume weight

Q3 A girl starts on the equator, walks 500 m East, 500 m South, 500 m West and 500 m North. State

a the distance she walked
b the magnitude of her final displacement.

Q4 Point B is 2000 km in a direction 30° East of North from Point A.

a To the nearest kilometre, how far is B North of A?
b How far is A West of B?

ANSWERS

1 Scalars and vectors

1 A scalar is a quantity that has size (magnitude) only.
A vector is a quantity that has both magnitude and direction.

2 Scalars: distance, mass, pressure, volume.
Vectors: acceleration, force, momentum, weight.

3 a The girl walked 2000 m (2·0 km).
  b Her final displacement is 0 m. (She finishes back where she started).

4 a B is 1732 km North of A.
  b A is 1000 km West of B.

***Exam* tip:** Make sure you understand the difference between scalars and vectors and that you know which quantities are scalars and which quantities are vectors – questions on this are common.

Higher Physics
Mechanics and properties of matter

2

Vector components

Q1 A boy kicks a football with an initial velocity of 10 m s⁻¹ at an angle of 30° to the horizontal. State the relationship you should use to calculate

  a the initial vertical component of the velocity of the ball
  b the initial horizontal component of the velocity of the ball.

Q2 A crate of mass m kg is at rest on a slope which is at an angle of $\theta°$ to the horizontal. State the relationship you should use to calculate the component of the crate's weight acting

  a perpendicular to the slope
  b parallel to the slope.

Q3 A force of 500 N is applied to a crate at an angle of 20° to the horizontal, as shown below.

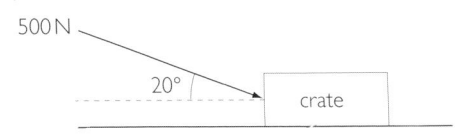

  a Calculate the horizontal component of the force.
  b Calculate the vertical component of the force.

ANSWERS

2 Vector components

1. a Vertical component = $10 \sin 30°$
   b Horizontal component = $10 \cos 30°$

2. a Component perpendicular to the slope = $mg \cos \theta°$
   b Component parallel to the slope = $mg \sin \theta°$

3. a Horizontal component = $500 \cos 20° = 470 \text{ N}$
   b Vertical component = $500 \sin 20° = 170 \text{ N}$

***Exam* tip:** For calculating vertical and horizontal components of a vector *always* use the angle between the vector and the horizontal. For components into and down slopes *always* use the angle between the slope and the horizontal.

Higher Physics
Mechanics and properties of matter

3

Adding vectors

Q1 What is meant by the term *resultant* of a number of forces?

Q2 Two forces of 150N are acting on an object.
  a State the maximum possible value of the resultant of these two forces.
  b State the minimum possible value of the resultant of these two forces.
  c Calculate the magnitude of the resultant force acting on the object when the angle between the directions of the two forces is 90°.
  d State the direction of the resultant of the forces in part **c**.

Q3 An aeroplane initially flying due North in still air at a speed of 110 m s^{-1} enters an area where the air is moving 25·0 m s^{-1} due East. Calculate the velocity of the aeroplane in the moving air.

Q4 State two methods for solving problems on the addition of vectors.

ANSWERS »

3 Adding vectors

1. The resultant of a number of forces is the *vector sum* of the forces.

2. **a** 300 N (both forces in the same direction).
   b 0 N (the forces in opposite directions).
   c $F_{resultant}^2 = 150^2 + 150^2$ (using Pythagoras' Theorem)
   $\Rightarrow F_{resultant} = 212$ N
   d The direction of the resultant force is at 45° to the direction of each force.

3. (speed of aeroplane)$^2 = 110^2 + 25^2$
   $\Rightarrow$ speed $= 113 \, \text{m s}^{-1}$
   $\tan \theta = \dfrac{25}{110}$
   $\Rightarrow \theta = 12 \cdot 8°$
   $\Rightarrow$ velocity of aeroplane $= 113 \, \text{m s}^{-1}$ 12·8° East of North.

4. By scale drawings or by using geometrical relationships (e.g. Pythagoras' Theorem, sine, cosine and tangent relationships).

***Exam* tip:** If you are asked to calculate a vector state *both direction and magnitude* in your final answer.

ANSWERS   *See page 4 re answer layout*

Higher Physics
Mechanics and properties of matter

4

Acceleration

Q1   **a**   Define *acceleration* of an object.
      b   State the quantities which must be measured in an experiment to find the acceleration of an object.

Q2   Name the quantities in the relationship below and state the SI unit of each.

$$a = \frac{v - u}{t}$$

Q3   What causes an object to accelerate?

Q4   What is the difference between positive acceleration and negative acceleration?

ANSWERS

4 Acceleration

1. **a** Acceleration of an object is the change in velocity per unit time.
   b Initial velocity, final velocity and time for this velocity change.

2. a – acceleration – metre per second per second (ms^{-2})
   v – final velocity – metre per second (ms^{-1})
   u – initial velocity – metre per second (ms^{-1})
   t – time – second (s)

3. An unbalanced force acting on an object causes it to accelerate.

4. Positive and negative accelerations are in opposite directions.

***Exam* tip:** When using the relationship $a = \dfrac{v - u}{t}$ make sure that you substitute the values of u and v in the correct places. Substituting values the wrong way round is a common error. It is wrong physics and costs marks.

***Exam* tip:** It is *incorrect physics* to say 'Positive acceleration is rate of speeding up and negative acceleration is rate of slowing down'. The direction of an acceleration is the same as the direction of the unbalanced force causing it. In straight line vector questions choose a positive direction – any vector in the opposite direction is negative.

Higher Physics
Mechanics and properties of matter

Graphs of motion

Q1 Sketch the displacement–time, velocity–time and acceleration–time graphs of an object moving with uniform velocity.

Q2 Sketch the displacement–time, velocity–time and acceleration–time graphs of an object moving with uniform acceleration from rest.

Q3 You have been given a velocity–time graph for a moving object.
   a How do you calculate the acceleration of the object?
   b How do you calculate the displacement of the object?

Q4 An object with an initial velocity of $5 \cdot 0 \, \text{ms}^{-1}$ accelerates uniformly to $10 \, \text{ms}^{-1}$ in a time of $10 \, \text{s}$. Calculate the distance travelled by the object during its acceleration.

ANSWERS

5 Graphs of motion

1

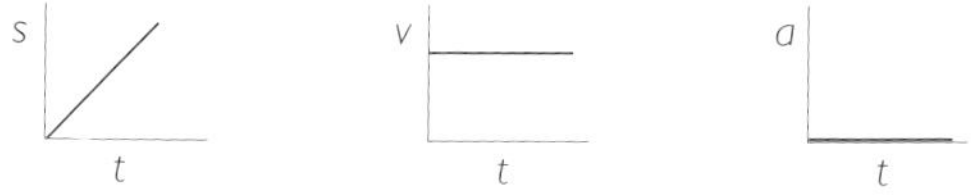

2

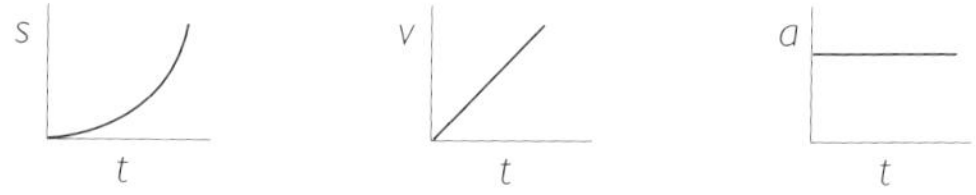

3 a Calculate the gradient of the graph – the acceleration is equal to the gradient.
  b Calculate the area between the graph and the time axis – the displacement is equal to this area.

4

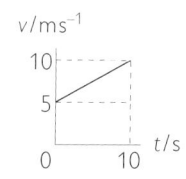

Area of rectangle = $5.0 \times 10 = 50$
Area of triangle = $½ \times 5.0 \times 10 = 25$
Distance travelled = $75 \, m$

Exam tip: Be very careful when you are answering questions on velocity–time graphs. Areas *above* the time axis are *positive*; areas *below* the time axis are *negative*.

Higher Physics
Mechanics and properties of matter

The equations of motion

Q1 Derive the first equation of motion.

Q2 Derive the second equation of motion.

Q3 Derive the third equation of motion.

Q4 State the *two* conditions which must be met so that the equations of motion may be used to solve a problem.

ANSWERS

6 The equations of motion

1 $a = \dfrac{v - u}{t}$

$\Rightarrow at = v - u$

$\Rightarrow v = u + at$

2

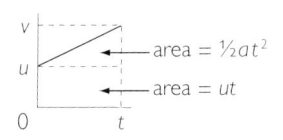

```
v |   
  |         /|
  |        / | ← area = ½at²
u |-------/  |
  |      |   | ← area = ut
  |_____|___|
  0      t
```

Distance = area under speed–time graph
area of rectangle = ut
area of triangle = $½ \times (v - u) \times t = ½at^2$
$\Rightarrow \quad s = ut + ½at^2$

3 $v = u + at$

$\Rightarrow v^2 = u^2 + 2uat + a^2t^2$

$\Rightarrow v^2 = u^2 + 2a(ut + ½at^2)$

$\Rightarrow v^2 = u^2 + 2as$

4 The equations of motion may only be used for objects which move with *uniform acceleration* in a *straight line*.

Exam tip: The conditions in Answer 4 are very important – make sure you do not try to use equations of motion for any other type of motion – that is *incorrect physics* and costs marks.

Higher Physics
Mechanics and properties of matter

7

Using the equations of motion

Q1 An advert for a car states that the car can accelerate from $0\,\mathrm{m\,s^{-1}}$ to $45\,\mathrm{m\,s^{-1}}$ in $7\cdot5\,\mathrm{s}$. Calculate this acceleration of the car.

Q2 An object dropped from a stationary hot air balloon falls to Earth in a time of $2\cdot3\,\mathrm{s}$.

  a Calculate the height of the balloon when the object was dropped.
  b State an assumption you have made in your calculation.

Q3 A speed boat travelling at $10\,\mathrm{m\,s^{-1}}$ accelerates uniformly at $0\cdot75\,\mathrm{m\,s^{-2}}$ for a distance of $200\,\mathrm{m}$. Calculate the final velocity of the boat.

Q4 An object dropped from a height of $28\cdot8\,\mathrm{m}$ falls to the surface of the Moon. The gravitational field strength on the Moon is approximately one-sixth the value of the gravitational field on Earth.

  a Calculate the gravitational field strength on the Moon.
  b Calculate the time for the object to fall to the surface of the Moon.

ANSWERS

7 Using the equations of motion

1 $u = 0 \, \text{ms}^{-1}$ $v = 45 \, \text{ms}^{-1}$ $t = 7 \cdot 5 \, \text{s}$

$$a = \frac{v - u}{t}$$
$$= \frac{45 - 0}{7 \cdot 5}$$
$$= 6 \cdot 0 \, \text{ms}^{-2}$$

2 a $t = 2 \cdot 3 \, \text{s}$ $u = 0 \, \text{ms}^{-1}$ $a = 9 \cdot 8 \, \text{ms}^{-2}$

$$s = ut + \tfrac{1}{2}at^2$$
$$= 0 + \tfrac{1}{2} \times 9 \cdot 8 \times 2 \cdot 3^2$$
$$= 26 \, \text{m}$$

  b The effect of air resistance is negligible.

3 $u = 10 \, \text{ms}^{-1}$ $a = 0 \cdot 75 \, \text{ms}^{-2}$ $s = 200 \, \text{m}$

$$v^2 = u^2 + 2as$$
$$= 100 + 2 \times 0 \cdot 75 \times 200$$
$$\Rightarrow v = 20 \, \text{ms}^{-1}$$

4 a $g_{\text{Moon}} = \dfrac{9 \cdot 8}{6} = 1 \cdot 6 \, \text{Nkg}^{-1}$

  b $u = 0 \, \text{ms}^{-1}$ $s = 28 \cdot 8 \, \text{m}$ $a = 1 \cdot 6 \, \text{ms}^{-2}$

$$s = ut + \tfrac{1}{2}at^2$$
$$\Rightarrow 28 \cdot 8 = 0 + \tfrac{1}{2} \times 1 \cdot 6 \times t^2$$
$$\Rightarrow t = 6 \cdot 0 \, \text{s}$$

***Exam* tip:** Most motion questions can be solved using only one equation. Solving a problem using more than one equation may take longer but it is not wrong and you could still get full marks.

* See page 4 re answer layout.

Higher Physics
Mechanics and properties of matter

8

Projectiles

Q1
a. Describe the horizontal motion of a projectile when resistance to the motion is negligible.
b. Describe the vertical motion of a projectile when resistance to the motion is negligible.
c. In your exam you should always consider the horizontal and vertical motions as if they happen separately. Why?

Q2
Simon drops a stone into a well and measures a time of 3·0 s between dropping the stone and hearing the splash when the stone hits the water. Simon tells Sarah that this means the well is 44 m deep. Sarah disagrees.

a. Write out Simon's calculation.
b. Who is correct, Simon or Sarah?

Q3
From a point level with the centre of a target, an archer fires an arrow at 25 m s^{-1} at an angle of 8·0° above horizontal. The arrow strikes the centre of the target.

a. Calculate the vertical component of the initial velocity of the arrow.
b. Hence calculate the time for the arrow to reach its highest point
c. Calculate the horizontal component of the initial velocity of the arrow.
d. Hence calculate the distance between the archer and the target.

ANSWERS

8 Projectiles

1 a Uniform velocity.

b Uniform acceleration.

c Considering the horizontal and vertical motions separately allows use of equations of motion without the method being incorrect physics.

2 a $u = 0\,\text{ms}^{-1}$ $t = 3\cdot0\,\text{s}$ $a = 9\cdot8\,\text{ms}^{-2}$

$s = ut + \tfrac{1}{2}at^2$

$= (0 \times 3\cdot0) + (\tfrac{1}{2} \times 9\cdot8 \times 3\cdot0^2)$

$= 44\,\text{m}$

b Sarah is correct. Simon has calculated the depth to the water surface.

Depth of the well $= 44\,\text{m} +$ depth of water.

3 a Vertical component of velocity $= 25\sin 8° = 3\cdot5\,\text{ms}^{-1}$

b $u = 3\cdot5\,\text{ms}^{-1}$ $v = 0\,\text{ms}^{-1}$ $a = -9\cdot8\,\text{ms}^{-2}$

$$a = \frac{v - u}{t}$$

$\Rightarrow \quad -9\cdot8 = \dfrac{0 - 3\cdot5}{t}$

$\Rightarrow$ time to highest point $= 0\cdot36\,\text{s}$

c Horizontal component of velocity $= 25\cos 8°$

$= 25\,\text{ms}^{-1}$

d Time of flight $= 2 \times 0\cdot36\,\text{s} = 0\cdot72\,\text{s}$

Horizontal distance to target $= vt$

$= 25 \times 0\cdot72 = 18\,\text{m}$

Exam tip: For every projectile the *time* for its horizontal motion is the same as the *time* for its vertical motion.

ANSWERS   * See page 4 re answer layout.

Higher Physics
Mechanics and properties of matter

9

Balanced and unbalanced forces

Q1   **a**   What is meant by the term *balanced* forces?
      b   What is meant by the term *unbalanced* forces?

Q2   **a**   How does an object move when the forces acting on it are balanced?
      b   How does an object move when the forces acting on it are unbalanced?

Q3   **a**   State the resultant force acting on an object falling vertically at its terminal velocity.
      b   Justify your answer.
      c   Can falling objects on the Moon reach a terminal velocity?

Q4   How can you tell when an unbalanced force is acting on an object?

Q5   Action–reaction forces are equal and opposite; why do they *not* cancel each other?

ANSWERS

9 Balanced and unbalanced forces

1 a Balanced forces act on the same object and cancel each other out – overall the force is 0N.

 b When the resultant force acting on an object is not zero the forces are unbalanced.

2 a The object remains at rest *or* travels with a constant velocity (i.e. at constant speed in a straight line).

 b The object accelerates (speed may increase *or* speed may decrease *or* direction may change).

3 a Resultant force = 0N.

 b The object has constant velocity (is moving with constant speed in a straight line)

 ⇒ forces acting on the object are balanced.

 c No – the Moon has no atmosphere – there is no air or any other gas to cause an upward force that could balance the downward force of the weight.

4 Observe the motion of the object – if it is changing speed or direction then there is an unbalanced force acting.

5 Action–reaction forces act on different objects. Equal and opposite forces can only cancel each other if they act on the same object.

***Exam* tip:** Understanding balanced and unbalanced forces is very important for explaining the way in which objects move – identifying whether forces are balanced or unbalanced is very useful for understanding questions on motion.

Higher Physics
Mechanics and properties of matter

10

Force, mass and acceleration

Q1 Define the *newton*.

Q2 What effect does a constant unbalanced force have on an object?

Q3 Choose the correct phrase to complete the following sentence:

When an unbalanced force acts on an object, the direction of the acceleration is…

the same as   *perpendicular to*   *opposite to*

…the direction of the unbalanced force.

Q4 State in words the relationship between unbalanced force, mass and acceleration.

Q5 Name the quantities in the relationship below and state the SI unit of each.

$F = ma$

ANSWERS

10 Force, mass and acceleration

1. One newton is the unbalanced force which causes an acceleration of $1\,\text{m s}^{-2}$ when it acts on a mass of $1\,\text{kg}$.

2. A constant unbalanced force causes an object to have a constant acceleration.

3. When an unbalanced force acts on an object, the direction of the acceleration is *the same as* the direction of the unbalanced force.

4. When an unbalanced force acts on an object the acceleration is directly proportional to the unbalanced force and inversely proportional to the mass of the object.

5. F – (unbalanced) force – newton (N)
   m – mass – kilogram (kg)
   a – acceleration – metre per second per second (m s^{-2})

***Exam* tip:** When using the relationship $F = ma$ remember that the quantity F is the *unbalanced force*.

Higher Physics
Mechanics and properties of matter

11

Analysing forces

Q1 An aeroplane is flying horizontally at a constant speed.

   a Name and state the directions of all of the horizontal forces acting on the aeroplane.
   b State the relationship between the magnitudes of the horizontal forces.
   c Name and state the directions all of the vertical forces acting on the aeroplane.
   d State the relationship between the magnitudes of the vertical forces.

Q2 A submarine is diving at a constant speed at an angle of 5° below horizontal.

   a State the direction of any unbalanced force acting on the submarine.
   b Justify your answer.

Q3 A woman is standing in a stationary lift in a department store. The lift is suspended in a lift shaft and is supported by a wire cable.

   a Name and state the directions of all of the vertical forces acting on the lift.
   b Name and state the directions of all of the vertical forces acting on the woman.
   c **i** State whether the vertical forces on the woman are balanced or unbalanced when the lift is accelerating upwards.
      ii State the relationship between the vertical forces acting on the woman when the lift is accelerating upwards.
      iii State the relationship between the vertical forces acting on the woman when the lift is accelerating downwards.

ANSWERS

11 Analysing forces

1 a Forward thrust of the engine(s), backward force of air resistance.
  b Forward thrust of engines = air resistance.
  c Upward buoyancy force, upthrust from wings, weight downwards.
  d Weight = buoyancy force + upthrust from wings.

2 a Forces acting on the submarine are balanced.
  b The submarine is moving in a straight line at a constant speed.

3 a Down: weight of lift, weight of woman; up: tension in cable.
  b Down: weight of woman; up: force from floor of the lift.
  c **i** Unbalanced.
     ii Force from floor > weight of woman.
     iii Force from floor < weight of woman.

***Exam* tip:** When you are analysing forces in more than one dimension always consider the forces *one dimension at a time*.

Higher Physics
Mechanics and properties of matter

12

Energy and work

Q1 Name the quantities in the relationships below and state the SI unit of each.
  a $E_k = \tfrac{1}{2}mv^2$
  b $E_p = mgh$
  c $E_w = Fd$

Q2 a State the energy changes in a swinging pendulum.
  b The bob of a pendulum is pulled to the side and released from a point 0·010 m higher than its lowest point. Calculate the speed of the bob at its lowest point.

Q3 A speed boat is initially stationary. The boat engine then exerts a constant force of 120 N through a distance of 300 m.
  a Calculate the work done by the boat engine.
  b State the maximum possible value of the final kinetic energy of the boat.
  c Why is the actual value of the final kinetic energy likely to be less than this value?

Q4 a Define *power*.
  b State whether it is a scalar or vector quantity.
  c State the SI unit of power.

ANSWERS

12 Energy and work

1. **a** E_k – kinetic energy – joule (J)
   m – mass – kilogram (kg)
   v – velocity – metre per second (m s^{-1})
   b E_p – gravitational potential energy – joule (J)
   m – mass – kilogram (kg)
   g – acceleration due to gravity – metre per second per second (m s^{-2})
   or gravitational field strength – newton per kilogram (N kg^{-1})
   h – height – metre (m)
   c E_W – work done – joule (J)
   F – force – newton (N)
   d – distance – metre (m)

2. **a** Kinetic energy changes to gravitational potential energy and back to kinetic energy.
   b $g = 9\cdot8$ m s^{-2} $h = 0\cdot010$ m
   $\frac{1}{2}mv^2 = mgh$
   $\Rightarrow v^2 = 2 \times 9\cdot8 \times 0\cdot010$
   $\Rightarrow v = 0\cdot44$ m s^{-1}

3. **a** $F = 120$ N $d = 300$ m
   $E_W = Fd$
   $= 120 \times 300$
   $= 36\,000$ J (36 kJ)
   b Maximum possible value of $E_k = 36$ kJ
   c Some of the work done by the engine is converted to other forms of energy, e.g. heat.

4. **a** Power is the rate at which energy is used *or* the rate at which work is done.
   b scalar
   c watt (W)

***Exam* tip:** Note that changes in energy can be used to solve problems on motion where objects do not have constant acceleration and/or they do not move in straight lines.

Higher Physics
Mechanics and properties of matter

Momentum

Q1
a. Define *momentum*.
b. State whether momentum is a scalar or vector quantity.
c. State the SI unit of momentum.

Q2 State the law of conservation of linear momentum.

Q3 The momentum of a car travelling at $20\,\text{m s}^{-1}$ is $2.2 \times 10^4\,\text{kg m s}^{-1}$.

a. Calculate the mass of the car.
b. Hence calculate the kinetic energy of the car.

Q4 Show that when two objects m_1 and m_2 interact in the absence of external forces the change in momentum of one is equal and opposite to the change in momentum of the other.

ANSWERS

13 Momentum

1 a The momentum of an object is the product of its mass and its velocity.

  b Vector – *so you need to remember about positive and negative directions.*

  c $kg\,m\,s^{-1}$

2 Provided there are no external forces, the total momentum before a collision or explosion is equal to the total momentum after the collision or explosion.

3 a $v = 20\,m\,s^{-1}$ $p = 2.2 \times 10^4\,kg\,m\,s^{-1}$

$$p = mv$$
$$\Rightarrow\quad 2.2 \times 10^4 = m \times 20$$
$$\Rightarrow\quad m = 1100\,kg$$

  b $m = 1100\,kg$ $v = 20\,m\,s^{-1}$

$$E_k = \tfrac{1}{2}mv^2$$
$$= \tfrac{1}{2} \times 1100 \times 20^2$$
$$= 220\,000\,J\ (220\,kJ)$$

4 Before collision: momentum of $m_1 = m_1u_1$ and momentum of $m_2 = m_2u_2$
After collision: momentum of $m_1 = m_1v_1$ and momentum of $m_2 = m_2v_2$
total momentum after collision = total momentum before collision

$$\Rightarrow\quad m_1v_1 + m_2v_2 = m_1u_1 + m_2u_2$$
$$\Rightarrow\quad m_1v_1 - m_1u_1 = -m_2v_2 + m_2u_2$$
$$\Rightarrow\quad m_1v_1 - m_1u_1 = -(m_2v_2 - m_2u_2)$$
$$\Rightarrow\quad \text{change in momentum of } m_1 = -(\text{change in momentum of } m_2)$$

***Exam* tip:** Get into the habit of sketching the situation described in a momentum question; use arrows to show the directions of velocities and momenta and write data on your diagram. This will help you make sense of the data.

* See page 4 re answer layout.

Higher Physics
Mechanics and properties of matter

14

Impulse

Q1
a. Define *impulse*.
b. State whether impulse is a scalar or vector quantity.
c. State the SI unit of impulse.

Q2 An object of mass m has initial velocity u and final velocity v. Show that the change in momentum of this mass is equal to the impulse due to the force acting on the object.

Q3 The force–time graph for a stationary ball being hit by a bat is shown below.

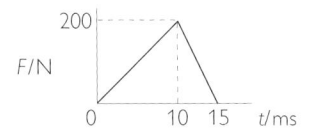

F/N — 200 at peak, 0 to 10 to 15 $t/$ms

a. Calculate the impulse on the ball.
b. State the change in momentum of the ball.
c. Calculate the average force acting on the ball.

ANSWERS

14 Impulse

1 a The impulse due to a force is the product of the force and the length of time the force acts.

b vector

c newton second (N s)

2 From Newton's second law, $F = ma = m\dfrac{v-u}{t}$

$$= \dfrac{mv - mu}{t}$$

$$\Rightarrow \quad Ft = mv - mu$$

3 Impulse = area under F–t graph
 = ½ × 200 × 15 × 10⁻³
 = 1·5 N s

b Change in momentum of ball = $1\cdot5\,\text{kg m s}^{-1}$

c Average force acting on ball = $\dfrac{\text{impulse}}{\text{total time}}$

$$= \dfrac{1\cdot5}{15 \times 10^{-3}} = 100\,\text{N}$$

***Exam* tip:** Did you notice that in **Q3** the time axis is labelled in milliseconds (ms)? Be careful when reading data from graphs – make sure you pay attention to the units on the axes.

ANSWERS

Higher Physics
Mechanics and properties of matter

Density

Q1
a. Define *density* of a material.
b. State whether density is a scalar or vector quantity.
c. State the SI unit of density.

Q2 A brass statue has mass 234 kg. The volume of the statue is 0·027 m³.

a. Calculate the density of the brass.
b. A cube of the same brass has the same mass as the statue. Calculate the length of the side of the cube.

Q3 Data on the density of selected solids, liquids and gases are shown in the table below.

Solid	Density/ kg m^{-3}	Liquid	Density/ kg m^{-3}	Gas	Density/ kg m^{-3}
aluminium	2700	water	1000	air	1·29
graphite	2300	glycerine	1260	nitrogen	1·25
amber	1100	olive oil	920	propane	2·02

a. How do the densities of gases compare to the densities of solids and the liquids?
b. Explain this observation in terms of particles.

Q4 Describe the principles of a method for measuring the density of air.

ANSWERS

15 Density

1 a Density of a material is the mass per unit volume.
  b scalar
  c kilogram per cubic metre (kg m^{-3})

2 a $m = 234$ kg $V = 0·027$ m^3

$$\rho = \frac{m}{V}$$
$$= \frac{234}{0·027}$$
$$= 8700 \text{ kg m}^{-3}$$

  b $V = 0·027$ m^3
  $\Rightarrow$ length of side of cube $= \sqrt[3]{0·027} = 0·30$ m

3 a The densities of the gases are of the order of 1000 times smaller.
  b Particles in gases are about ten times further apart than particles in solids or liquids.

4 Use a very sensitive balance and an airtight container. Weigh the container empty and again when it is full of air. Note both measurements and calculate the difference to find the mass of the air.

Measure the volume of the container when it is full of air.

Divide the mass of air by the volume to calculate its density.

***Exam* tip:** If you are asked a question on density, remember that the densities of some common materials are listed on the data page at the start of the Higher Physics paper.

ANSWERS * See page 4 re answer layout.

Higher Physics
Mechanics and properties of matter

16

Pressure

Q1
a. Define *pressure*.
b. State whether pressure is a scalar or vector quantity.
c. State the SI unit of pressure.

Q2 A man of mass 94 kg is standing upright on both feet. The area in contact with the floor of the sole plus the heel of each of his shoes is 330 cm^2.

a. Calculate the weight of the man.
b. Calculate the pressure exerted on the floor by the man's shoes.
c. State the pressure that is exerted on the floor when the man stands upright on one foot. Explain your answer.

Q3 A cube of aluminium of side 125 mm rests on a horizontal desk. Calculate the pressure exerted by the aluminium on the desk. (You may use data from the table in the previous topic.)

ANSWERS

16 Pressure

1 a Pressure is the force per unit area acting perpendicular to a surface.
   b scalar
   c pascal (Pa); $1\,\text{Pa} = 1\,\text{N}\,\text{m}^{-2}$

2 a $W = mg$
$= 94 \times 9\cdot 8$
$= 920\,\text{N}$

   b $A = 2 \times 330 \times 10^{-4}\,\text{m}^2$ $\quad F = W = 920\,\text{N}$

$P = \dfrac{F}{A}$

$= \dfrac{920}{0\cdot 066}$

$= 1\cdot 4 \times 10^4\,\text{Pa}$

   c $P = 2\cdot 8 \times 10^4\,\text{Pa}$

The man's weight is the same and the area supporting his weight is half.

3 $V = 0\cdot 125^3 = 1\cdot 95 \times 10^{-3}\,\text{m}^3$ $\quad \rho_{\text{aluminium}} = 2700\,\text{kg}\,\text{m}^{-3}$

$m = \rho V$
$= 2700 \times 1\cdot 95 \times 10^{-3}\,\text{kg}$

$F = W = 2700 \times 1\cdot 95 \times 10^{-3} \times 9\cdot 8\,\text{N} = 51\cdot 6\,\text{N}$

$A_{\text{cube face}} = 0\cdot 125^2 = 0\cdot 0156\,\text{m}^2$

$P = \dfrac{F}{A}$

$= \dfrac{51\cdot 6}{0\cdot 0156}$

$= 3300\,\text{Pa}$

***Exam* tip:** In the definition of pressure it is important to state that force is *perpendicular to the surface* – without this the definition is incomplete and you will lose marks.

Higher Physics
Mechanics and properties of matter

17

Pressure in fluids

Q1 What does the term *fluid* mean?

Q2
a What causes the pressure at a point in a fluid?
b What are the factors which affect the pressure in a fluid?
c State the relationship for calculating the pressure at a point in a fluid.

Q3 Pressure can cause force and force is a vector. How can pressure be a scalar?

Q4
a State the approximate value of air pressure at the surface of the Earth.
b Give two reasons why this value not constant.

Q5 Calculate the pressure at a depth of 25 m below the surface of the sea. Density of sea water $= 1.02 \times 10^3 \, kg \, m^{-3}$.

ANSWERS

17 Pressure in fluids

1. A fluid is a material which flows. (*Gases and liquids are fluids.*)

2. a Pressure at a point in a fluid is caused by the weight of fluid(s) above that point.
   b Pressure in a fluid is affected by depth, density of the fluid and gravitational field strength.
   c $P = g\rho h$

3. Pressure in a fluid acts in all directions so can cause force in any direction. The direction of force is perpendicular to any surface placed in the fluid.

4. a $1 \cdot 01 \times 10^5$ Pa
   b Air pressure changes with the weather and with height above sea level.

5. $g = 9 \cdot 8 \, \text{ms}^{-2}$ $\rho = 1 \cdot 02 \times 10^3 \, \text{kgm}^{-3}$
   $P = g\rho h + \textit{atmospheric pressure}$
   $= (9 \cdot 8 \times 1 \cdot 02 \times 10^3 \times 25) + 1 \cdot 01 \times 10^5$
   $= 3 \cdot 5 \times 10^5 \, \text{Pa}$

***Exam* tip:** On Earth the pressure at a depth in lochs, lakes or the sea is due to both atmosphere and water – this is easily forgotten during an exam.

Higher Physics
Mechanics and properties of matter

18

Buoyancy force

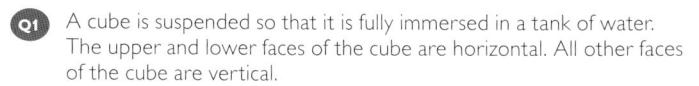

Q1 A cube is suspended so that it is fully immersed in a tank of water. The upper and lower faces of the cube are horizontal. All other faces of the cube are vertical.

  a Explain why there is a net upward force exerted on the cube by the water.
  b Explain why there is no net horizontal force on the cube.

Q2 Why is the buoyancy force on an object immersed in air much smaller than the buoyancy force on the same object immersed in water?

Q3 A diving bell has a uniform cross-sectional area of $20\,m^2$ and a height of $3.0\,m$. The bell is suspended in a freshwater loch so that the upper surface is at a depth of $5.0\,m$ below the surface of the loch.
Density of water $= 1.00 \times 10^3\,kg\,m^{-3}$.

  a Calculate the buoyancy force acting on the diving bell.
  b Calculate the volume of the diving bell.
  c State the volume of water displaced by the diving bell.
  d Hence calculate the weight of water displaced by the diving bell.
  e Compare your answers to parts **a** and **d**.

ANSWERS

18 Buoyancy force

1 a The area of the upper face is equal to the area of the lower face.

The lower face is at a greater depth than the upper face.

The water density and gravitational field strength are both constant.

⇒ upward pressure on the lower face > downward pressure on the upper face

⇒ force due to the water pressure on the lower face > the force due to the water pressure on the upper face

⇒ the water exerts a net upward force on the cube.

b Areas of vertical faces of the cube are equal.

All faces have the same pressure at the same depth.

⇒ forces due to water pressure on opposite vertical faces are equal and opposite

⇒ no net horizontal force on the cube.

2 The density of water is much greater than the density of air.

3 a $A = 20\,\text{m}^2$ $\quad g = 9.8\,\text{m s}^{-2}$ $\quad \rho = 1.00 \times 10^3\,\text{kg m}^{-3}$ $\quad (h_2 - h_1) = 3.0\,\text{m}$

Pressure difference $= \dfrac{F_{\text{upward}} - F_{\text{downward}}}{A} = g\rho(h_2 - h_1)$

Net upward force $= A \times$ pressure difference

⇒ Buoyancy force $F = 20 \times 9.8 \times 1.00 \times 10^3 \times 3.0 = 5.9 \times 10^5\,\text{N}$

b $A = 20\,\text{m}^2$ $\quad h = 3.0\,\text{m}$ $\quad V_{\text{diving bell}} = Ah = 20 \times 3.0 = 60\,\text{m}^3$

c $V_{\text{displaced water}} = 60\,\text{m}^3$

d $\rho_{\text{water}} = 1.00 \times 10^3\,\text{kg m}^{-3}$ $\quad g = 9.8\,\text{m s}^{-2}$

$W = mg = V\rho g$

$W = 60 \times 1.00 \times 10^3 \times 9.8 = 5.9 \times 10^5\,\text{N}$

e The answers to parts **a** and **d** are the same.

(Upthrust = weight of fluid displaced is Archimedes' Principle)

***Exam* tip:** In the Higher exam *never* call the upward force due to a fluid 'buoyancy' alone – always use the term *buoyancy force* or *upthrust*.

Higher Physics
Mechanics and properties of matter

19

Temperature

Q1
a State the relationship for converting from degrees celsius to kelvin.
b What is the difference between 1 degree celsius and 1 celsius degree?
c Compare the size of 1 celsius degree to 1 K.

Q2 Convert the following temperatures to degrees celsius.
a 273 K
b 100 K
c 0 K
d 400 K

Q3 Explain the difference between *temperature* and *heat*.

Q4 What is the meaning of the term *absolute zero* of temperature?

ANSWERS

19 Temperature

1 a $T_K = T_C + 273$

b 1 °C is a specific temperature just above the freezing point of water. 1 celsius degree is a change in temperature of 1 degree – it could be any change of 1 degree, for example from 20 °C to 21 °C, or 21 °C to 22 °C, etc.

c 1 celsius degree = 1 K

2 a 0 °C

b −173 °C

c −273 °C

d 127 °C

3 Heat is a form of energy. Temperature is the degree of hotness of a material.

(Heat is the sum of potential energy and kinetic energy of the particles; temperature in kelvin is proportional to the average kinetic energy of the particles.)

4 Absolute zero is the zero of the kelvin temperature scale. It is the lowest possible temperature in the universe. It is not possible to get any temperature lower than 0 K.

(At a temperature of 0 K both the kinetic energy and potential energy of the particles are zero.)

***Exam* tip:** If you get confused when converting between kelvin and degrees celsius remember temperatures in *kelvin* are *always bigger* than temperatures in degrees celsius; also, temperatures in *kelvin* are *never negative*.

42 ANSWERS

Higher Physics
Mechanics and properties of matter

20

Kinetic model of matter and gas pressure

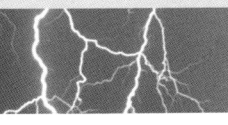

Q1 What is the meaning of the term *kinetic model of matter*?

Q2
  a Describe the structure of solids in terms of the kinetic model.
  b Describe the structure of liquids in terms of the kinetic model.
  c Describe the structure of gases in terms of the kinetic model.

Q3 Why does a solid have a definite shape yet a liquid takes up the shape of any container in which it is placed?

Q4 Solid iron melts to form liquid iron at 1539 °C.
  a What happens to the average kinetic energy of the iron atoms?
  b What happens to the average spacing of the iron atoms?

Q5
  a Explain gas pressure in terms of the kinetic model.
  b In the kinetic model, what factors affect the pressure produced by gas particles?

ANSWERS

20 Kinetic model of matter and gas pressure

1 The kinetic model of matter is a model used to explain the behaviour of materials in terms of the forces between particles and the movement of particles.

2 **a** In a solid the particles are very close together and are held in place by very strong forces. The particles in solids vibrate from side to side; they do not move around.

   b In liquids the particles are very close together and are held together by forces weaker than the forces in solids. The particles in liquids move randomly inside the liquid and frequently collide with neighbouring particles.

   c In gases the particles are about ten times further apart than the particles in solids and liquids. The forces between gas particles are very weak. The particles move randomly and frequently collide with other particles.

3 Solids have a definite shape because the particles are not able to move about; each particle has a set location in the solid. Liquids take up the shape of a container because the particles move about within the liquid.

4 **a** The average kinetic energy of the iron atoms remains the same.

   b The average spacing of the iron atoms remains the same.

5 **a** Gas pressure on a surface is due to gas particles colliding with the surface.

   b The pressure produced by gas particle collisions on a surface depends on:
   - the average force with which particles hit the surface – this depends on the mass and the average speed of the particles
   - the number of particle collisions with the surface per second.

***Exam* tip:** In Higher Physics, if you are asked to explain gas pressure, you must refer to the number of collisions *per second*. If you only mention the number of collisions you will lose marks.

Higher Physics
Mechanics and properties of matter

21

Gas laws

You may assume air pressure = 1.01×10^5 Pa

Q1
a State the pressure–volume gas law.
b Name the quantities in the relationship $P_1V_1 = P_2V_2$ and state the SI unit of each.
c Write down the alternative version of this relationship.
d A bicycle pump has an internal volume of 120 ml when the plunger is fully extended. A girl seals the end of the pump with her thumb and pushes the plunger until the internal volume is 25 ml. Calculate the final pressure of the air.
e In the calculation for part **d** why is it acceptable to use units which are not the SI unit for volume?

Q2
a State the pressure–temperature gas law.
b Write down a mathematical relationship based on this law.
c At the start of a climb a mountaineer is equipped with a cylinder of oxygen at a temperature of 20 °C and a pressure of 1.2×10^6 Pa. She climbs to a point where the temperature is −3 °C.
   i Calculate the new pressure of the gas.
   ii State two assumptions you have made in your calculation.

Q3 State the temperature–volume gas law.

ANSWERS

21 Gas laws

1 a The pressure of a fixed mass of gas at constant temperature is inversely proportional to its volume.

b P_1 – initial pressure – pascal (Pa); V_1 – initial volume – cubic metre (m³)
P_2 – final pressure – pascal (Pa); V_2 – final volume – cubic metre (m³)

c $PV = constant$

d $P_1 = 1.01 \times 10^5 \text{Pa} \quad V_1 = 120\,\text{ml} \quad V_2 = 25\,\text{ml}$
$P_1V_1 = P_2V_2 \Rightarrow 1.01 \times 10^5 \times 120 = P_2 \times 25 \Rightarrow P_2 = 4.8 \times 10^5 \text{Pa}$

e This is acceptable provided the volume units of the values substituted are the same on both sides of the relationship (the volume units cancel each other).

2 a The pressure of a fixed mass of gas at constant volume is directly proportional to its temperature measured in kelvin.

b $\dfrac{P}{T} = constant \quad or \quad \dfrac{P_1}{T_1} = \dfrac{P_2}{T_2}$

c i $T_1 = 20°C = 293\,\text{K} \quad T_2 = -3°C = 270\,\text{K} \quad P_1 = 1.2 \times 10^6 \text{Pa}$

$\Rightarrow \dfrac{1.2 \times 10^6}{293} = \dfrac{P_2}{270} \Rightarrow P_2 = 1.1 \times 10^6 \text{Pa}$

ii Assumptions: volume of cylinder is constant; all oxygen in the cylinder is at the given temperature; mass of oxygen is constant (no oxygen escaped or was used).

3 The volume of a fixed mass of gas at constant pressure is directly proportional to its temperature measured in kelvin.

***Exam* tip**: If you are asked to state a gas law and you want to get full marks make sure you include the relevant conditions, i.e. constant mass and one of the following three – constant pressure *or* constant volume *or* constant temperature – depending on the law.

ANSWERS * See page 4 re answer layout.

Higher Physics
Mechanics and properties of matter

More gas laws and kinetic model of matter

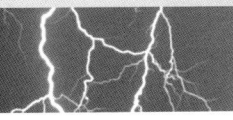

Q1 Explain the pressure–volume gas law in terms of kinetic theory.

Q2 Explain the pressure–temperature gas law in terms of kinetic theory.

Q3 Explain the volume–temperature gas law in terms of kinetic theory.

Q4 A gas cylinder of volume $0.800\,m^3$ is filled with nitrogen at a pressure of $1.80 \times 10^6\,Pa$ and a temperature of $10\,°C$. Calculate the maximum volume of nitrogen that can be supplied from this cylinder in a laboratory maintained at a constant $22\,°C$.

ANSWERS

22 More gas laws and kinetic model of matter

1 As the temperature is constant the average kinetic energy of the particles does not change and so the particles hit the container wall with the same average force. When the volume of the gas is increased the particles have to travel further on average between collisions with the container walls. There are fewer particle collisions per second with the walls. This causes the pressure to fall.

2 When temperature rises the average kinetic energy of particles increases and the particles move faster. The particles hit the walls with greater average force. Provided the volume does not change the particles also hit the walls more often each second. These two effects cause the pressure to increase.

3 As temperature falls the average kinetic energy of the particles decreases and the particles move more slowly. The particles hit the wall with less average force. To keep pressure constant the volume of the gas must decrease so that the particles hit the walls more often each second. So at constant pressure, when temperature decreases, volume also decreases.

4 $V_1 = 0.800 \, \text{m}^3$ $P_1 = 1.80 \times 10^6 \, \text{Pa}$ $T_1 = 283 \, \text{K}$
  $P_2 = 1.01 \times 10^5 \, \text{Pa}$ (P_2 is atmospheric pressure) $T_2 = 295 \, \text{K}$

$$\frac{P_1 V_1}{T_1} = \frac{P_2 V_2}{T_1}$$

$$\Rightarrow \frac{0.800 \times 1.8 \times 10^6}{283} = \frac{1.01 \times 10^5 \times V_2}{295}$$

$$\Rightarrow V_2 = 14.9 \, \text{m}^3$$

Maximum volume which can be supplied = $14.9 - 0.8 = 14.1 \, \text{m}^3$

***Exam* tip**: When explaining the pressure–temperature law or the volume–temperature law you may consider temperature increasing or decreasing – you do not need to include both.

ANSWERS   * See page 4 re answer layout.

Higher Physics
Electricity and electronics

Electric fields

Q1
a. Define the term *electric field*.
b. State the convention for the direction of an electric field.

Q2 State the direction of the force experienced by a negative charge placed in an electric field.

Q3 What happens when an electric field is applied to a conductor?

Q4 What happens when charge is moved in an electric field?

Q5 Copy and complete the following sentences:

'When _____ is moved in an _____ field the _____ done depends only on the _____ and _____ points. It does not depend on the _____ followed by the _____.'

ANSWERS

23 Electric fields

1. **a** An electric field is a region of space where a charge experiences an electrical force.
   b The direction of an electric field is the direction of the force experienced by a positive charge placed in the field.

2. A negative charge experiences a force in the opposite direction to the direction of an electric field.

3. An electric field applied to a conductor causes the free electric charges in it to move.

4. When charge is moved in an electric field work is done.

5. When *charge* is moved in an *electric* field the *work* done depends only on the *start* and *finish* points. It does not depend on the *path* followed by the *charge*.

Exam tip: Read carefully any question that you get on electric fields – be particularly careful to make sure you get the sign(s) of the charge(s) *and* the direction of the field correct.

Higher Physics
Electricity and electronics

Charge, work and potential difference

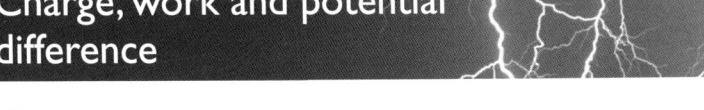

Q1 Define the term *potential difference* between two points.

Q2 Copy and complete the following sentence:
'If one joule of _____ is done moving one coulomb of _____ between two points, the _____ difference between the two points is one _____.'

Q3 Name the quantities in the relationship below and state the SI unit of each.

$$W = QV$$

Q4 Calculate the energy needed to move a charge of 3.00×10^{-6} C through a potential difference of 255 V.

24 Charge, work and potential difference

1. The potential difference between two points is the work done in moving one coulomb of charge between the two points.

2. If one joule of *work* is done moving one coulomb of *charge* between two points, the *potential* difference between the two points is one *volt*.

3. W – work done – joule (J)
   Q – charge – coulomb (C)
   V – potential difference – volt (V)

4. $Q = 3.00 \times 10^{-6}$ C $\qquad V = 255$ V

   $W = QV$
   $ = 3.00 \times 10^{-6} \times 255$
   $ = 7.65 \times 10^{-4}$ J

***Exam* tip:** Potential difference between two points is an important and fundamental concept in the physics of electricity – it is numerically equal to the electrical potential energy difference for each coulomb of charge which moves between these points.

Higher Physics
Electricity and electronics

E.m.f. and terminal potential difference

Q1 Define the term *e.m.f.* of a source of electrical energy.

Q2 Define the term *terminal potential difference* of a source of electrical energy.

Q3 When are the terminal potential difference and e.m.f. of source of electrical energy equal?

Q4 What is the meaning of the term *lost volts*?

ANSWERS

25 E.m.f. and terminal potential difference

1 The e.m.f. of a source of electrical energy is the electrical potential energy given to each coulomb of charge which passes through the source.

2 The terminal potential difference is the potential difference between the terminals of a source of electrical energy.

3 Terminal potential difference and e.m.f. of a source of electrical energy are equal when the current in the source is zero.

4 'Lost volts' is the difference between the e.m.f. and the terminal p.d. of a source of electrical energy.

   or 'Lost volts' is equal to the energy required to move one coulomb of charge through the source.

***Exam* tip:** A source of electrical energy is an electrical machine and like any other machine it is not 100% efficient. Energy lost in driving charge through the source accounts for the lost volts.

Higher Physics
Electricity and electronics

E.m.f. and internal resistance

Q1 Copy and complete the following sentence:

'A source of electrical _____ is equivalent to an _____ in _____ with a small _____, the internal _____.'

Q2 Name the quantities in the relationship below and state the SI unit of each.

$$E = V + Ir$$

Q3 A battery of e.m.f. 12 V is connected in series with a 45 Ω resistor. The current in the resistor is 0·25 A. Calculate the internal resistance of the battery.

Q4 Describe the principles of a method for measuring the e.m.f. and internal resistance of a source of electrical energy.

ANSWERS

26 E.m.f. and internal resistance

1 A source of electrical *energy* is equivalent to an *e.m.f.* in *series* with a small *resistor*, the internal *resistance*.

2 E – e.m.f. – volt (V)
V – terminal potential difference – volt (V)
I – current – amp (A)
r – internal resistance – ohm (Ω)

3 $E = 12\,\text{V}$ $R = 45\,\Omega$ $I = 0{\cdot}25\,\text{A}$
$$V_{resistor} = IR$$
$$= 0{\cdot}25 \times 45$$
$$E = V + Ir$$
$$\Rightarrow 12 = (0{\cdot}25 \times 45) + 0{\cdot}25r$$
$$\Rightarrow r = 3{\cdot}0\,\Omega$$

4 Connect the source of electrical energy in series with a variable resistor and an ammeter and in parallel with a voltmeter.

Note the voltmeter and ammeter readings.

Adjust the variable resistor and again take both meter readings. Repeat until at least five pairs of readings have been obtained.

Plot a graph of voltmeter reading (*y*-axis in volts) *vs* ammeter reading (*x*-axis in amps); a straight line graph is obtained.

Extend the graph until it cuts the *y*-axis – the e.m.f. is equal to voltage value at this point.

Calculate the gradient of the straight line; the magnitude of the gradient is equal to the internal resistance of the source.

Exam tip: The concept of internal resistance helps us to quantify the energy lost driving charge through a source of electrical energy.

Higher Physics
Electricity and electronics

27

Energy in electrical circuits

Q1 In an electrical circuit how does charge gain electrical potential energy?

Q2 An electrical circuit has three d.c. electrical sources of e.m.f. E_1, E_2 and E_3 connected in series with a load resistor R. Adjacent sources have terminals connected positive to negative.

   a Write down an expression for the total energy gained by charge Q moving through the three sources.

   b The source with e.m.f. E_2 is reversed. Write down an expression for the total energy now gained by charge Q moving through the three sources.

Q3 **a** Where in an electrical circuit does charge lose electrical potential energy?

   b Give one example of a circuit situation in which charge loses electrical potential energy, and state what happens to the electrical energy lost.

Q4 An electrical circuit has three resistors R_1, R_2 and R_3 connected in series with an e.m.f. E. The potential differences across the resistors are V_1, V_2 and V_3.

   a Write down an expression for the total energy lost by charge Q moving through the three potential differences.

   b Resistor R_2 is reversed. Write down an expression for the total energy now lost by charge Q moving through the three potential differences.

Q5 State the relationship between the total energy gained and the total energy lost by charge moving round the circuit.

ANSWERS

27 Energy in electrical circuits

1. Charge gains electrical potential energy when it passes through a source of e.m.f.

2. **a** Total energy gained = $Q \times (E_1 + E_2 + E_3)$
   b Total energy gained = $Q \times (E_1 - E_2 + E_3)$

3. **a** Charge loses electrical potential energy when it passes through a circuit component in which electrical energy is converted to another form of energy.
   b Resistor: electrical energy $\longrightarrow$ heat
   Motor: electrical energy $\longrightarrow$ kinetic energy
   LED: electrical energy $\longrightarrow$ light; etc.

4. **a** Total energy lost = $Q \times (V_1 + V_2 + V_3)$
   b Total energy lost = $Q \times (V_1 + V_2 + V_3)$

5. Total energy gained = total energy lost.

Exam tip: Energy is an important quantity in all aspects of physics – for this reason energy questions often integrate ideas and concepts from more than unit of the Higher course.

Higher Physics
Electricity and electronics

28

Power in electrical circuits

Q1 Define the *power* in an electrical circuit.

Q2 Name the quantities in the relationship below and state the SI unit of each.

$$P = \frac{W}{t}$$

Q3 Starting from the above relationship show that $P = IV$, where the symbols have their usual meanings.

Q4 An electrical supply of e.m.f. 25 V and internal resistance $2 \cdot 0\,\Omega$ is connected to a load resistor of $48\,\Omega$.

  a Calculate the current drawn from the supply.
  b Calculate the power dissipated in the load resistor.
  c Calculate the power wasted.

ANSWERS

28 Power in electrical circuits

1. Electrical power is electrical work done per second.

2. P – power – watt (W)
   W – work or energy – joule (J)
   t – time – second (s)

3. $P = \dfrac{W}{t}$

   $W = QV \quad \Rightarrow \quad P = \dfrac{QV}{t}$
   $$= \dfrac{Q}{t} \times V$$
   $$= IV$$

4. **a** $E = 25\,\text{V} \qquad r = 2\cdot0\,\Omega \qquad R = 48\,\Omega$
   $E = I(R + r)$
   $\Rightarrow 25 = I \times (48 + 2\cdot0)$
   $\Rightarrow I = 0\cdot50\,\text{A}$

   b $I = 0\cdot50\,\text{A} \qquad R = 48\,\Omega$
   $P = I^2 R$
   $= 0\cdot50^2 \times 48$
   $= 12\,\text{W}$

   c Power wasted in $r = P = I^2 r$
   $= 0\cdot50^2 \times 2\cdot0$
   $= 0\cdot50\,\text{W}$

Exam tip. Every year many physics candidates forget to square the value of current when they are using $P = I^2 r$. Make sure this does not happen to you.

Higher Physics
Electricity and electronics

29

Current, voltage and resistance in series circuits

Q1 State the rule for current in a series circuit.

Q2 State the rule for potential difference in a series circuit.

Q3 By consideration of conservation of energy, derive the expression for the total resistance of any number of resistors in series.

Q4 Three resistors of resistance $10\,\Omega$, $12\,\Omega$ and $14\,\Omega$ are connected in series with a $9\cdot0\,V$ d.c. supply.

  a Calculate the total circuit resistance.
  b Calculate the current in the circuit.
  c Calculate the p.d. across the $12\,\Omega$ resistor.

ANSWERS ▶

29 Current, voltage and resistance in series circuits

1 In a series circuit current is the same at all points.

2 In a series circuit p.d.s add up to the p.d. across the source of electrical energy.

3 A battery of voltage V is connected in series to three resistors of resistance R_1, R_2 and R_3. Let the total resistance of the circuit be R and the current in the circuit be I.

Consider charge Q going round the circuit.

Energy gained by Q going through battery $= QV = QIR$
(using $V = IR$)
Energy lost by Q going through $R_1 = QV_1 = QIR_1$
Energy lost by Q going through $R_2 = QV_2 = QIR_2$
Energy lost by Q going through $R_3 = QV_3 = QIR_3$
Total energy gained $=$ total energy lost

$\Rightarrow \quad QIR = QIR_1 + QIR_2 + QIR_3$
$\Rightarrow \quad R = R_1 + R_2 + R_3$

4 a $R_1 = 10\,\Omega \quad R_2 = 12\,\Omega \quad R_3 = 14\,\Omega$
$R = R_1 + R_2 + R_3$
$= 10 + 12 + 14 = 36\,\Omega$

b $V = 9\cdot0\,\text{V} \quad R = 36\,\Omega$
$V = IR$
$\Rightarrow \quad 9\cdot0 = I \times 36$
$\Rightarrow \quad I = 0\cdot25\,\text{A}$

c $R_2 = 12\,\Omega \quad I = 0\cdot25\,\text{A}$
$V = IR_2$
$= 0\cdot25 \times 12 = 3\cdot0\,\text{V}$

Exam tip: In a series circuit, the total resistance is *always bigger* than the biggest single resistance.

ANSWERS   * See page 4 re answer layout.

Higher Physics
Electricity and electronics

Current, voltage and resistance in parallel circuits

Q1 State the rule for current in a parallel circuit.

Q2 State the rule for potential difference in a parallel circuit.

Q3 By consideration of conservation of charge, derive the expression for the total resistance of any number of resistors in parallel.

Q4 Three resistors of resistance $3.0\,\Omega$, $9.0\,\Omega$ and $18\,\Omega$ are connected in parallel with a $9.0\,\text{V}$ d.c. supply.
   a Calculate the total circuit resistance.
   b Calculate the current in the circuit.
   c Calculate the current in the $9.0\,\Omega$ resistor.

ANSWERS

30 Current, voltage and resistance in parallel circuits

1. In a parallel circuit currents in parallel branches add up to the current in the source of electrical energy.

2. Potential differences across parallel branches are the same as the p.d. across the source of electrical energy.

3. A battery of voltage V is connected in parallel to three resistors of resistance R_1, R_2 and R_3. Let the total resistance of the circuit be R and the current from the battery be I. Let the currents in R_1, R_2 and R_3 be I_1, I_2 and I_3 respectively.

$$V = IR \quad \Rightarrow I = \frac{V}{R}$$

p.d across each resistor $= V$

$$\Rightarrow I_1 = \frac{V}{R_1} \text{ and } I_2 = \frac{V}{R_2} \text{ and } I_3 = \frac{V}{R_3}$$

As charge in conserved $I = I_1 + I_2 + I_3$

Substituting $\Rightarrow \dfrac{V}{R} = \dfrac{V}{R_1} + \dfrac{V}{R_2} + \dfrac{V}{R_3} \Rightarrow \dfrac{1}{R} = \dfrac{1}{R_1} + \dfrac{1}{R_2} + \dfrac{1}{R_3}$

4. **a** $R_1 = 3.0\,\Omega \quad R_2 = 9.0\,\Omega \quad R_3 = 18\,\Omega$

$$\frac{1}{R} = \frac{1}{R_1} + \frac{1}{R_2} + \frac{1}{R_3} = \frac{1}{3.0} + \frac{1}{9.0} + \frac{1}{18} \Rightarrow R = 2.0\,\Omega$$

b $V = 9.0\,V \quad R = 2.0\,\Omega$

$$V = IR$$
$$\Rightarrow 9.0 = I \times 2.0 \quad \Rightarrow \quad I = 4.5\,A$$

c $R_2 = 9.0\,\Omega \quad V = 9.0\,V$

$$V = IR_2$$
$$\Rightarrow 9.0 = I \times 9.0 \quad \Rightarrow \quad I = 1.0\,A$$

Exam tip: In a parallel circuit, the total resistance is *always smaller* than the smallest single resistance.

ANSWERS   * See page 4 re answer layout.

Higher Physics
Electricity and electronics

31

Current, voltage and resistance in more complex circuits

Q1 A 12·0 V d.c. supply is connected with three resistors, as shown in the circuit below. Each resistor has a resistance of 4·0 Ω.

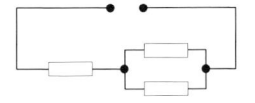

  a Calculate the total resistance of the circuit.
  b Calculate the current from the d.c. supply.
  c State the current in each parallel branch. Justify your answer.
  d Calculate the p.d. across the parallel branches.
  e State the p.d. across the resistor in series. Justify your answer.

Q2 A 3·0 V d.c. supply is connected with four resistors, as shown in the circuit below.

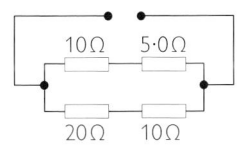

  a Calculate resistance of the upper parallel branch.
  b Calculate the current in the upper parallel branch.
  c Calculate resistance of the lower parallel branch.
  d Calculate the current in the lower parallel branch.
  e State the current from the d.c. supply. Justify your answer.

ANSWERS

31 Current, voltage and resistance in more complex circuits

1 a $R_1 = R_2 = 4 \cdot 0 \, \Omega$

For the parallel resistors $\dfrac{1}{R} = \dfrac{1}{R_1} + \dfrac{1}{R_2} = \dfrac{1}{4 \cdot 0} + \dfrac{1}{4 \cdot 0} \Rightarrow R = 2 \cdot 0 \, \Omega$

Total resistance $R_{total} = R_3 + R = 4 \cdot 0 + 2 \cdot 0 = 6 \cdot 0 \, \Omega$

b $V = 12 \cdot 0 \, V \quad R_{total} = 6 \cdot 0 \, \Omega$

$V = IR_{total} \Rightarrow 12 \cdot 0 = I \times 6 \cdot 0 \Rightarrow I = 2 \cdot 0 \, A$

c Current in each parallel branch $= 1 \cdot 0 \, A$

The resistance of and p.d. across each branch are equal.

d $R_1 = R_2 = 4 \cdot 0 \, \Omega \qquad I = 1 \cdot 0 \, A$

$V = IR_2 = 1 \cdot 0 \times 4 \cdot 0 = 4 \cdot 0 \, V$

e $V = 8 \cdot 0 \, V$

P.d. across
series resistor = d.c. supply voltage − p.d. across parallel branches
$\qquad \qquad \qquad = 12 \cdot 0 - 4 \cdot 0 = 8 \cdot 0 \, V$

2 a $R_1 = 10 \, \Omega \qquad R_2 = 5 \cdot 0 \, \Omega$

$R = R_1 + R_2 = 10 + 5 \cdot 0 = 15 \, \Omega$

b $R = 15 \, \Omega \qquad V = 3 \cdot 0 \, V$

$V = IR \Rightarrow 3 \cdot 0 = I \times 15 \Rightarrow I = 0 \cdot 2 \, A$

c $R_3 = 20 \, \Omega \qquad R_4 = 10 \, \Omega$

$R = R_3 + R_4 = 20 + 10 = 30 \, \Omega$

d $R = 30 \, \Omega \qquad V = 3 \cdot 0 \, V$

$V = IR \Rightarrow 3 \cdot 0 = I \times 30 \Rightarrow I = 0 \cdot 1 \, A$

e Current from d.c. supply $= 0 \cdot 3 \, A$

$\qquad = $ sum of currents in parallel branches

Exam tip: It is likely that a circuit in your Higher paper will have both series and parallel parts. The best way to learn how to tackle these is to practise. The more you practise the better you will become.

66 **ANSWERS** * See page 4 re answer layout.

Higher Physics
Electricity and electronics

32

Potential divider

Though not an explicit requirement, understanding potential dividers will help you understand the Wheatstone bridge and questions on electronic circuits.

Q1 A 5·0V d.c. supply is connected to two resistors as shown in the diagram.

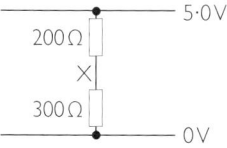

- 200 Ω
- X
- 300 Ω
- 5·0 V
- 0 V

   a Calculate the current in the circuit.
   b Calculate the p.d. across the 300 Ω resistor.
   c Hence state the voltage at point X.
   d The resistors are interchanged. State the voltage at X now. Justify your answer.

Q2 A 12V d.c. supply is connected to four resistors as shown in the diagram.

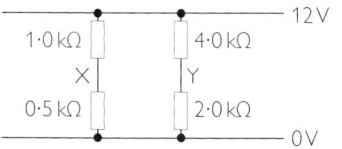

- 1·0 kΩ
- X
- 0·5 kΩ
- 4·0 kΩ
- Y
- 2·0 kΩ
- 12 V
- 0 V

   a Calculate the voltage at X.
   b Calculate the voltage at Y.
   c A high-resistance voltmeter is connected between X and Y. State the reading on the meter. Justify your answer.
   d **i** Calculate the ratio between the resistances in the X-branch.
      ii Calculate the ratio between the resistances in the Y-branch.

ANSWERS

32 Potential divider

1 a Total circuit resistance $= 200 + 300 = 500\,\Omega$.
Substituting in $V = IR \Rightarrow 5\cdot0 = I \times 500$
$\Rightarrow I = 0\cdot01\,\text{A}\ (10\,\text{mA})$

b Substituting in $V = IR \Rightarrow V = 0\cdot01 \times 300 = 3\cdot0\,\text{V}$

c Voltage at $X = 3\cdot0\,\text{V}$.

d Voltage at $X = 2\cdot0\,\text{V}$. As nothing else has changed in the circuit, the p.d. across the $300\,\Omega$ resistor is still $3\cdot0\,\text{V} \Rightarrow$ p.d. at $X = (5\cdot0 - 3\cdot0)\,\text{V}$.

2 a Total resistance in the X-branch $= (R_1 + R_2) = 1\cdot5\,\text{k}\Omega$.
$V = 12\,\text{V}$
Substituting in $V = IR \Rightarrow 12 = I \times 1\cdot5 \times 10^3$
Substituting for I in $V = IR_2 \Rightarrow V_X = \dfrac{12}{1\cdot5 \times 10^3} \times 0\cdot5 \times 10^3$
$\Rightarrow V_X = 4\cdot0\,\text{V}$

b Similarly $V_Y = \dfrac{12}{6\cdot0 \times 10^3} \times 2\cdot0 \times 10^3 = 4\cdot0\,\text{V}$.

c Reading on voltmeter $= 0\,\text{V}$.
Voltages at X and Y are the same $\Rightarrow$ p.d. between X and Y $= 0\,\text{V}$.

d i Ratio $= \dfrac{R_1}{R_2} = \dfrac{1\cdot0}{0\cdot5} = 2$

ii Ratio $= \dfrac{R_3}{R_4} = \dfrac{4\cdot0}{2\cdot0} = 2$

Exam tip: You may find these relationships useful:

for a potential divider

$$\text{p.d. across } R_1 = V \times \dfrac{R_1}{R_1 + R_2} \qquad \text{and} \qquad \text{p.d across } R_2 = V \times \dfrac{R_2}{R_1 + R_2}$$

Higher Physics
Electricity and electronics

33

Wheatstone bridge

Q1   **a**   Sketch and label a Wheatstone bridge circuit.
       b   Now identify two pairs of circuit components from your labelled circuit which form two potential dividers.

Q2   Under what conditions is a Wheatstone bridge *balanced*?

Q3   Copy and complete the following:

'For an initially _____ Wheatstone bridge, as the value of one _____ is _____ by a _____ amount, the _____ _____ is _____ proportional to the change in _____.'

Q4   In a balanced Wheatstone bridge R_1 and R_2 form a potential divider, and R_3 and R_4 form another potential divider. The values of the resistance of three resistors are as follows:

$R_1 = 10\,000\,\Omega$, $R_2 = 15\,000\,\Omega$, $R_3 = 8000\,\Omega$.

   a   Calculate the value of R_4.
   b R_4 is replaced with a variable resistor and the bridge is again balanced. The variable resistance is increased by $100\,\Omega$; the reading on the voltmeter is noted as $2\cdot5\,\text{mV}$. The variable resistance is increased by a further $300\,\Omega$. Calculate the reading on the voltmeter now.

ANSWERS »

33 Wheatstone bridge

1 a

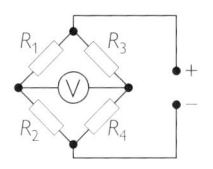

[Circuit diagram showing a Wheatstone bridge with resistors R_1, R_2, R_3, R_4 and a voltmeter V, connected to + and − terminals.]

b Resistors R_1 and R_2; and resistors R_3 and R_4.

2 A Wheatstone bridge is balanced when the p.d. measured by the voltmeter is zero;

or when $\dfrac{R_1}{R_2} = \dfrac{R_3}{R_4}$.

3 For an initially *balanced* Wheatstone bridge, as the value of one *resistor* is *changed* by a *small* amount, the *out-of-balance p.d.* is *directly* proportional to the change in *resistance*.

4 a Substituting in $\dfrac{R_1}{R_2} = \dfrac{R_3}{R_4}$ $\Rightarrow$ $\dfrac{10\,000}{15\,000} = \dfrac{8000}{R_4}$

$\Rightarrow$ $R_4 = 12\,000\,\Omega$

b Total increase in variable resistance $= (100 + 300) = 400\,\Omega$

New reading on voltmeter $= 2 \cdot 5 \times \dfrac{400}{100} = 10\,\text{mV}$

***Exam* tip:** A Wheatstone bridge circuit can be drawn either as shown above in the answer to **Q1a** or as the diagram in **Q2** of Topic 32.

Higher Physics
Electricity and electronics

34

Alternating current and voltage

Q1 The mains electricity supply in the UK has a declared r.m.s. voltage of 230 V and a frequency of 50 Hz.
  a Calculate the peak voltage of the mains supply.
  b Calculate the time for 1 complete cycle of the mains supply.
  c How may times each second does the mains voltage reach its peak value?

Q2 Describe how to measure the frequency of an alternating voltage using an oscilloscope.

Q3 A signal generator is connected in series with a load resistor and an ammeter. The output voltage of the signal generator is initially set to 12 V and 40 Hz. The frequency of the output voltage is gradually increased; the voltage is maintained at 12 V. Describe and explain what happens to the reading on the ammeter.

Q4 An a.c. current in a 370 Ω resistor has a peak value of 0·36 A. Calculate the power dissipated in the resistor.

ANSWERS

34 Alternating current and voltage

1 a $V_{peak} = \sqrt{2} \times V_{r.m.s.} = \sqrt{2} \times 230 = 325\,\text{V}$

b Period $= \dfrac{1}{\text{frequency}} = \dfrac{1}{50} = 0.020\,\text{s}$

c Peak voltage occurs twice each cycle (1 positive and 1 negative) $\Rightarrow$ peak value occurs 100 times each second.

2 Connect the supply to the input terminals of a calibrated oscilloscope.

Adjust the time-base setting of the oscilloscope to give a suitable number of waves on the screen.

Calculate the number of divisions on the oscilloscope screen for one complete wave and multiply this by the time-base setting to obtain the period T of the a.c. supply.

Use the relationship $f = \dfrac{1}{T}$ to find the frequency of the supply.

3 The reading on the ammeter remains constant. In a circuit containing only resistors, current is independent of frequency; when frequency is changed, current remains the same.

4 Substituting in $I_{peak} = \sqrt{2} \times I_{r.m.s.}$ $\Rightarrow$ $0.36 = \sqrt{2} \times I_{r.m.s.}$
$\Rightarrow$ $I_{r.m.s.} = 0.255\,\text{A}$

Power dissipated $= I_{r.m.s.}^{2} \times R = 0.255^{2} \times 370 = 24\,\text{W}$

***Exam* tip:** When calculating power or energy in an a.c. circuit *always* use r.m.s. values of current and voltage.

Higher Physics
Electricity and electronics

35

Energy and power in resistive circuits

Q1 A plasma television draws an r.m.s. current of 1·50 A from the mains supply. The television is on for an average of 8 hours every day.
  a Calculate the power used by the television.
  b Calculate the energy in kWh consumed by the television in a period of 4 weeks.
  c Calculate the energy in joules used in one day.

Q2 A battery of e.m.f. 9·00 V is connected in series with a resistor of 5·00 Ω and a combination of a 3·00 Ω resistor and 6·00 Ω resistor in parallel. The current in the 3·00 Ω resistor is 0·800 A.
  a State the current in the 6·00 Ω resistor. Justify your answer.
  b State the current in the 5·00 Ω resistor.
  c Calculate the total power dissipated in the resistors.
  d Calculate the power supplied by the e.m.f.
  e Account for any difference between the answers to parts **c** and **d**.

Q3 A voltage of 20 V is applied to a 40 Ω resistor. Calculate the power dissipated in the resistor.

ANSWERS

35 Energy and power in resistive circuits

1 a Power $P = IV \Rightarrow P = 1.50 \times 230 = 345\,\text{W}$

b 4 weeks = 28 days
$\Rightarrow$ total energy used $= 0.345 \times 8 \times 28 = 77.3\,\text{kWh}$

c $P = 345\,\text{W}$ 8 hours $= 8 \times 60 \times 60 = 28\,800\,\text{s}$
$E = Pt = 345 \times 28\,800 = 9.94\,\text{MJ}$ ($9.94 \times 10^6\,\text{J}$)

2 a Current in $6.00\,\Omega$ resistor $= 0.400\,\text{A}$.
p.d. across the $6\,\Omega$ = p.d. across the $3\,\Omega$;
resistance double $\Rightarrow$ current half

b Current in $5.00\,\Omega$ resistor $= 1.20\,\text{A}$.

c $5.00\,\Omega$ resistor:
Substituting in $P = I^2R \Rightarrow P = 1.2^2 \times 5.0 = 7.20\,\text{W}$
$3.00\,\Omega$ resistor:
Substituting in $P = I^2R \Rightarrow P = 0.8^2 \times 3.0 = 1.92\,\text{W}$
$6.00\,\Omega$ resistor:
Substituting in $P = I^2R \Rightarrow P = 0.4^2 \times 6.0 = 0.96\,\text{W}$
Total power dissipated $= 7.20 + 1.92 + 0.96 = 10.1\,\text{W}$

d Substituting in $P = IV \Rightarrow P = 1.20 \times 9.00 = 10.8\,\text{W}$

e Power (0.7 W) is wasted in the internal resistance of the battery.

3 Power $P = \dfrac{V^2}{R} \Rightarrow P = \dfrac{20^2}{40} = 10\,\text{W}$

***Exam* tip:** There are three relationships for calculating power in electrical circuits:

$$P = IV \qquad P = I^2R \qquad P = \dfrac{V^2}{R}$$

Similarly there are three relationships for calculating energy:

$$E = IVt \qquad E = I^2R\,t \qquad E = \dfrac{V^2}{R}t$$

Choose the right version and you can save time.

Higher Physics
Electricity and electronics

Capacitance 1

Q1 Describe the structure of a capacitor.

Q2 State the relationship between the charge Q on two parallel conducting plates and the p.d. V between the plates.

Q3
a Define the quantity *capacitance*.
b State the SI unit of capacitance.
c Define the SI unit of capacitance.

Q4
a Can a capacitor become fully charged in *any* electrical circuit?
b When is a capacitor fully charged?

Q5 A 12 V battery is connected to a 400 µF capacitor in series with a 300 Ω resistor.

a Calculate the charge on the capacitor when it is fully charged.
b The capacitor is discharged and the 300 Ω resistor is replaced by a 500 Ω resistor. What effect does this have on the charge on the capacitor when it is fully charged?
c The capacitor is discharged and the battery is replaced by a 15 V battery. Calculate the charge now on the capacitor when it is fully charged.

ANSWERS

36 Capacitance 1

1 A capacitor consists of two parallel conducting plates separated by an insulating material.

2 The charge Q on two parallel conducting plates is directly proportional to the p.d. V between the plates.

3 a Capacitance is the ratio of charge to p.d. $C = \dfrac{Q}{V}$.
  b farad (F)
  c One farad is one coulomb per volt; $1\,\text{F} = 1\,\text{C}\,\text{V}^{-1}$.

4 a No, a capacitor can only become fully charged when it is connected to a d.c. source of electrical energy.
  b A capacitor is fully charged when it reaches the p.d. of the source.

5 a Substituting in $C = \dfrac{Q}{V}$ $\Rightarrow$ $400 \times 10^{-6} = \dfrac{Q}{12}$
  $\Rightarrow$ $Q = 4\cdot8 \times 10^{-3}\,\text{C}$
  b Changing the resistor has no effect on the charge on the fully charged capacitor.
  c Substituting in $C = \dfrac{Q}{V}$ $\Rightarrow$ $400 \times 10^{-6} = \dfrac{Q}{15}$
  $\Rightarrow$ $Q = 6\cdot0 \times 10^{-3}\,\text{C}$

***Exam* tip:** The p.d. across a capacitor is in the opposite direction to the p.d. of the source to which it is connected. In a d.c. circuit current becomes zero when the net p.d. around the circuit is zero.

Higher Physics
Electricity and electronics

37

Capacitance 2

Q1 Explain why work must be done to charge a capacitor.

Q2
  a Describe the shape of the graph of charge against p.d. for a capacitor.
  b How can this graph be used to calculate the work done to charge a capacitor?
  c State the relationship between work done to charge a capacitor and the energy stored by the capacitor.

Q3 Name the quantities in the relationship below and state the SI unit of each.

$W = \tfrac{1}{2}QV$

Q4 A charge of $9 \cdot 0 \times 10^{-3}$ C is placed on the plates of a $1 \cdot 5$ mF capacitor.

  a Calculate the p.d. between the plates of the capacitor.
  b Calculate the energy stored by the capacitor.

ANSWERS

37 Capacitance 2

1 Consider two initially uncharged parallel plates A and B.

A negative charge is removed from A. A is positively charged. The negative charge is attracted to A. Work must be done to move the charge to B. The negative charge is placed on B. B is negatively charged.

A second negative charge is removed from A. Positive charge on A is bigger. The charge is more attracted to A and repelled by B. More work must be done to move the charge to B. The second charge is placed on B. Negative charge on B is bigger.

More work must be done for each charge moved from A to B.

2 a The graph is a straight line with positive slope, passing through the origin.

  b The work done to charge a capacitor is given by the area under the graph of charge against p.d.

  c Work done to charge a capacitor = energy stored by the capacitor.

3 W – work done to charge the capacitor – joule (J)
  Q – charge on the capacitor – coulomb (C)
  V – p.d. across the capacitor – volt (V)

4 a Substituting in $C = \dfrac{Q}{V}$ $\Rightarrow$ $1.5 \times 10^{-3} = \dfrac{9.0 \times 10^{-3}}{V}$

$\Rightarrow V = 6.0\,V$

  b Energy stored $E = \frac{1}{2}QV = \frac{1}{2} \times 9.0 \times 10^{-3} \times 6.0 = 0.027\,J$

Exam tip: There are three relationships for calculating energy stored by a capacitor:

$$E = \tfrac{1}{2}QV \qquad E = \tfrac{1}{2}CV^2 \qquad E = \tfrac{1}{2}\dfrac{Q^2}{C}$$

Choosing the right version can save you time during your exam.

ANSWERS

Higher Physics
Electricity and electronics

38

Capacitors in d.c. circuits

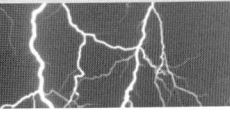

Q1
a Sketch a graph of current against time for charging a capacitor in a d.c. circuit containing a resistor and capacitor in series.
b Sketch a graph of p.d. against time for charging a capacitor in a d.c. circuit containing a resistor and capacitor in series.

Q2
a Sketch a graph of current against time for discharging a capacitor in a d.c. circuit containing a resistor and capacitor in series.
b Sketch a graph of p.d. against time for discharging a capacitor in a d.c. circuit containing a resistor and capacitor in series.
c Explain the significance of the shape of the graphs for **Q1a** and **Q2a**.

Q3 A 20 V battery is connected to a 5·0 mF capacitor in series with a 100 Ω resistor and an open switch. The switch is closed and the capacitor is allowed to become fully charged. The capacitor is discharged and the 100 Ω resistor is replaced by a 50 Ω resistor. The switch is closed and the capacitor is allowed to become fully charged again. What effect does changing the resistor have on the charging process? Give a reason for your answer.

Q4 By combining the relationships $V = IR$ and $C = \dfrac{Q}{V}$ and simplifying, find the SI units for the product RC.

ANSWERS »

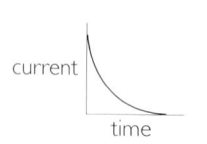

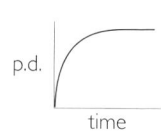

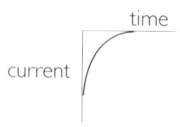

38 Capacitors in d.c. circuits

1 a [graph: current decreasing exponentially with time]

b [graph: p.d. rising exponentially with time]

2 a [graph: current rising exponentially with time]

b [graph: p.d. decreasing exponentially with time]

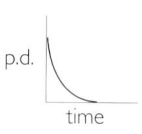

c The discharging current is in the opposite direction to the charging current.

3 With the smaller resistor the capacitor becomes fully charged more quickly. Initial current is bigger as the circuit resistance is smaller.

4 $R = \dfrac{V}{I} = \dfrac{Vt}{Q}$ $C = \dfrac{Q}{V}$

$\Rightarrow RC = \dfrac{Vt}{Q} \times \dfrac{Q}{V} = t$

$\Rightarrow$ SI unit for RC is second (s).

***Exam* tip:** In a circuit containing a resistor and a capacitor in series, the greater the value of the product RC, the longer it takes for the capacitor to charge or discharge. This idea may help you answer questions on charging and discharging of capacitors.

Higher Physics
Electricity and electronics

39

Capacitors in a.c. circuits

Q1   **a**   What is meant by the term *capacitive circuit*?
       b   State the relationship between current and frequency in a capacitive a.c. circuit.

Q2   An initially uncharged capacitor is connected to an a.c. supply. A steady current is obtained. When the same capacitor is connected to a d.c. supply the current quickly falls to zero. Explain this difference.

Q3   **a**   Sketch a circuit which can be used to show how the current varies with frequency in a capacitive circuit.
       b   Briefly describe how the circuit is used to show the relationship.

Q4   When an a.c. supply is connected in a resistive circuit energy is dissipated in resistors. When an a.c. supply is connected in a capacitive circuit no energy is dissipated in the capacitors. Explain.

ANSWERS

39 Capacitors in a.c. circuits

1 a A capacitive circuit is a circuit in which the only components are a source of electrical energy and capacitors (and meter(s)).

b In a circuit containing only capacitors, current is directly proportional to the frequency of the a.c. supply.

2 In the a.c. circuit the capacitor begins to charge during the first half cycle of the a.c. It discharges during the next half cycle. This process continues for each cycle of the a.c. so that the capacitor never becomes fully charged. In the d.c. circuit the capacitor quickly becomes fully charged.

3 a

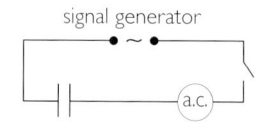

[Circuit diagram: signal generator connected to a capacitor and an a.c. ammeter with a switch]

b Close the switch. Record the frequency of the supply and the reading on the a.c. ammeter. Increase the frequency of the a.c supply. Again record the frequency and current readings. Repeat until at least five pairs of readings have been obtained. Plot a graph of *current vs frequency*. A straight line graph passing through the origin is obtained.

4 In the first half of an a.c. cycle, energy from the supply is stored in the electric field which is built up in the capacitor. In the second half of the a.c. cycle this energy is returned to the supply.

***Exam* tip:** A capacitor does not oppose current in the same way as a resistor. The p.d. built up on a capacitor acts against the e.m.f. of the supply and so reduces the net voltage in the circuit.

Higher Physics
Electricity and electronics

40

Uses of capacitors

Q1   **a**   Describe a practical application of using a capacitor to store energy.
      b   How is the energy stored by the capacitor?

Q2   **a**   Describe a practical application of using a capacitor to store charge.
      b   How is the charge stored by the capacitor?

Q3   **a**   Describe how a capacitor may be used to remove the d.c. part of a signal which is part a.c. and part d.c.
      b   How does the capacitor remove the d.c. part of the signal?

ANSWERS

40 Uses of capacitors

1 a Stored energy in a capacitor can be used to provide a short pulse of current in the flash unit of a camera. The capacitor is charged from the camera battery. When the shutter is pressed the capacitor discharges through the bulb giving an intense pulse of light.

b Energy is stored in the electric field between the plates of the capacitor.

2 a A capacitor connected in parallel with the output of a d.c. source reduces variation of the voltage of the output. The p.d. across the capacitor stabilises the voltage because time is needed to charge and discharge the capacitor.

b The charge is stored as an excess of positive charge on one plate and an equal excess of negative charge on the other.

3 a The d.c. part can be removed by passing the signal through a capacitor in series.

b The capacitor charges up to a p.d. equal and opposite to the p.d. of the d.c. part of the signal.

***Exam* tip:** Questions on capacitors may include part(s) set in the context of practical uses of capacitors. Make sure you are familiar with these uses.

Higher Physics
Electricity and electronics

Operational amplifiers

Q1 What is an operational amplifier (op-amp)?

Q2 Sketch and label the symbol for an op-amp.

Q3
a. What are the characteristics of an ideal op-amp?
b. How does the behaviour of real op-amps compare to an ideal op-amp?

Q4
a. Explain the term *saturation*.
b. When does an op-amp become saturated?

ANSWERS

41 Operational amplifiers

1 An operational amplifier is an electronic device that is used to increase the voltage of a signal.

2

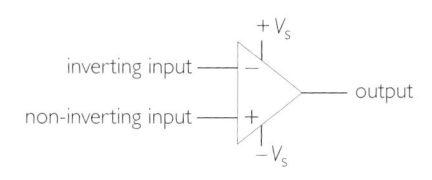

```
                              +V_s
                               |
   inverting input ———————|—\
                          |   >——— output
   non-inverting input ———|+/
                               |
                              -V_s
```

3 a For an ideal op-amp:
- input current is zero *or* the op-amp has infinite input resistance
- both inputs are at the same potential.

b The behaviour of real op-amps is very close to ideal.

4 a An op-amp cannot produce an output voltage greater than the positive supply voltage or lower than the negative supply voltage. Saturation occurs when it reaches its maximum or minimum voltage.

b When the magnitude of the output voltage reaches a value a little below the magnitude of the supply voltage the op-amp becomes saturated.

***Exam* tip:** It is *incorrect* to say that the voltage is saturated – if you write this you will lose marks – it is the op-amp which is saturated, not the voltage.

86 ANSWERS

Higher Physics
Electricity and electronics

Op-amps in circuits 1

Q1   **a**   Name the quantities in the relationship below and state the SI unit of each.
$$\frac{V_o}{V_1} = -\frac{R_f}{R_1}$$
    b   What is the significance of the negative sign in this relationship?

Q2   An ideal op-amp is connected in inverting mode to an input voltage of 0·01 V. The resistance of R_1 is 100 Ω. The output voltage is −2·0 V. Calculate the value of R_f.

Q3   An op-amp is connected in inverting mode in a circuit where R_1 is 2000 Ω and R_f is 10 kΩ. A 1·4 V r.m.s. 25 Hz sinusoidal a.c. signal is connected to the inverting input of the op-amp. The supply voltage of the op-amp is ±14 V.

    a   Calculate the peak value of the output voltage.
    b   The input and output are connected to an oscilloscope.
        i   Describe the shape of the trace of the output voltage.
        ii   Compare the traces of the output and input voltages.
    c   The signal is replaced with a 1·4 V r.m.s. 50 Hz sinusoidal source. What effects does this have on the amplitude and number of waves in the trace of the output?
    d   Resistor R_1 is replaced with a resistor of resistance 0·80 kΩ. Describe the shape of the trace of the output voltage.

ANSWERS

42 Op-amps in circuits 1

1 a V_o – output voltage – volt (V)
V_1 – input voltage – volt (V)
R_f – feedback resistance – ohm (Ω)
R_1 – resistance between input voltage and inverting input – ohm (Ω)

b The negative sign means that the input voltage is inverted – that is if the input is positive then the output is negative and vice versa.

2 $V_o = -2.0\,\text{V}$ $V_1 = 0.01\,\text{V}$ $R_1 = 100\,\Omega$

Substituting in $\dfrac{V_o}{V_1} = -\dfrac{R_f}{R_1} \Rightarrow \dfrac{-2.0}{0.01} = -\dfrac{R_f}{100}$

$\Rightarrow R_f = 20000\,\Omega = 20\,\text{k}\Omega$

3 a For the input voltage, $V_{peak} = \sqrt{2} \times V_{r.m.s.} = \sqrt{2} \times 1.4 = 2.0\,\text{V}$

Substituting in $\dfrac{V_o}{V_1} = -\dfrac{R_f}{R_1} \Rightarrow \dfrac{V_o}{2.0} = \dfrac{-10000}{2000}$

$\Rightarrow V_o = -10\,\text{V}$

b i The output voltage is a sine wave (sinusoidal).

ii Amplitude of output voltage is greater than amplitude of input voltage. The traces have the same number of waves.

c The amplitude of the waves is the same; the number of waves is doubled.

d The output trace is a sine wave with each crest and trough cut off at the voltage at which the op-amp becomes saturated.

$\dfrac{V_o}{2.0} = \dfrac{-10000}{800}$

$\Rightarrow V_o = -25\,\text{V}$

Exam tip: When you are tackling problems on electric/electronic circuits, sketch and label a circuit diagram and use your diagram to help you understand what is happening in the circuit.

ANSWERS   * See page 4 re answer layout.

Higher Physics
Electricity and electronics

Op-amps in circuits 2

Q1
a. Name the quantities in the relationship below and state the SI unit of each.

$$V_o = (V_2 - V_1)\frac{R_f}{R_1}$$

b. In what mode is an op-amp operating when this relationship is used?
c. How many components need to be added to complete the circuit? What are these components?
d. What is the relationship between the values of these additional components and components included in the above relationship?

Q2 The diagram shows a Wheatstone bridge connected to an op-amp in differential mode.

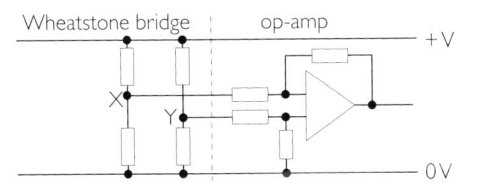

a. Give two examples of possible resistive sensors used in this type of circuit.
b. Where is the best location for a sensor in the Wheatstone bridge circuit?
c. Describe the function of the op-amp in this circuit.
d. Describe how the circuit might be used with a transistor and relay to switch on a high-power circuit.

ANSWERS

43 Op-amps in circuits 2

1 a V_o – output voltage – volt (V)
  V_1 – voltage connected to inverting input – volt (V)
  V_2 – voltage connected to non-inverting input – volt (V)
  R_f – resistance between the op-amp output and inverting input – ohm (Ω)
  R_1 – resistance between input voltage and inverting input – ohm (Ω)
 b Differential mode.
 c Two additional resistors are needed to complete the circuit:
  R_2 resistance between a second input voltage and the non-inverting input
  R_3 resistance between non-inverting input and Earth (zero volts).
 d Resistor values are chosen so that $\dfrac{R_f}{R_1} = \dfrac{R_3}{R_2}$.

2 a LDR, thermistor, strain gauge, etc.
 b The resistive sensor may be located in any one of the four resistor locations in the Wheatstone bridge.
 c The op-amp is measuring the difference in p.d. between points X and Y.
 d The output of the op-amp is connected to the base of the transistor and the relay is connected so that when the transistor is switched on the relay is switched on. The changing of the op-amp output voltage from negative to positive switches on the transistor and relay. The relay operates the high power switch.

Exam tip: The circuits like the one in **Q3** are complicated, so handle the Wheatstone bridge and op-amp calculations separately. The result of one gives information that you use in the other. The order in which you do these calculations depends on the question.

Higher Physics
Radiation and matter

Wave properties

Q1 What is a wave?

Q2
a. What does the term *medium* mean in relation to wave travel?
b. Name three types of wave and give one example of a medium for each.
c. Electromagnetic waves can pass through a vacuum where there is no material to vibrate and carry the waves. How is energy carried in electromagnetic waves?

Q3
a. What is a transverse wave?
b. Sketch a transverse wave. On your sketch indicate an amplitude and a wavelength.

Q4
a. What is a longitudinal wave?
b. In a longitudinal wave what terms are equivalent to 'crest' and 'trough' of a transverse wave?
c. Give two examples of longitudinal waves.

Q5 Name the quantities in the relationships below and state the SI unit of each.

a. $T = \dfrac{1}{f}$

b. $v = f\lambda$

ANSWERS

44 Wave properties

1. A wave is a regular oscillation that carries *energy* through a material or vacuum.

2. **a** A medium is the substance (or vacuum) through which a wave travels.
   b Visible light – vacuum between Sun and Earth.
   Sound waves – air.
   Water wave – surface of a loch or pond.
   c In electromagnetic waves energy is carried in the form of oscillating electric and magnetic fields of the same frequency.

3. **a** In a transverse wave the displacement of the medium is at right angles to the direction in which the wave is travelling.
   b

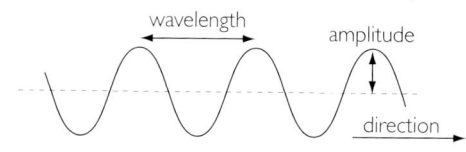

   [Diagram of a transverse wave showing wavelength between two crests, amplitude from midline to crest, and direction of travel along the horizontal axis.]

4. **a** In a longitudinal wave the displacement of the medium is parallel to the direction in which the wave is travelling.
   b Compression (area of higher pressure) and rarefaction (area of lower pressure).
   c Sound, ultrasound, forward and backward vibrations in a stretched spring, 'P' seismic waves.

5. **a** T – period of wave – second (s)
   f – frequency of wave – hertz (Hz)
   b v – wave speed – metre per second (m s^{-1})
   f – frequency of wave – hertz (Hz)
   λ – wavelength – metre (m)

Exam tip: If you find it difficult to visualise longitudinal waves try visualising them as transverse waves – the basic principles are the same for both.

Higher Physics
Radiation and matter

Wave definitions

Q1
a. Define the *amplitude* of a wave.
b. State the link between amplitude and energy carried by a wave.

Q2
a. Define the *frequency* of a wave.
b. What determines the frequency of a wave?

Q3
a. Define the *speed* of a wave.
b. What determines the speed of a wave?

Q4 Define the *period* of a wave.

Q5 Define the *wavelength* of a wave.

Q6
a. What does the term *in phase* mean?
b. What does the term *out of phase* mean?

Q7 What is meant by the term *coherent* sources of waves?

ANSWERS

45 Wave definitions

1 a Amplitude of a wave is the maximum displacement of the medium from its rest position (its position when there is no wave present).

  b The greater the amplitude, the greater the energy carried by a wave.

2 a Frequency of a wave is the number of waves made each second *or* the number of waves passing a point each second.

  b The frequency of a wave is determined by the source of the wave – the frequency of the wave is equal to the frequency of vibration of the source.

3 a The speed of a wave is the rate at which the wave travels through a medium.

  b The speed of a wave is determined by the medium.

4 The period of a wave is the time to make one complete wave.

5 Wavelength is the length of one complete wave measured parallel to the direction in which the wave is travelling.

6 a On a continuous wave, any two points a whole number of wavelengths apart are in phase. For example all crests are in phase with each other.

  b On a continuous wave, any two points an odd number of half wavelengths apart are out of phase. For example a crest and a trough are out of phase.

7 Two sources of waves are *coherent* when they have the same frequency and there is a constant phase difference between them.

***Exam* tip:** It is very important that you are comfortable using the language of waves and that you understand the basic physical quantities associated with waves *and* the factors which determine these quantities.

ANSWERS

Higher Physics
Radiation and matter

Wave behaviour

Q1
a. What is meant by the term *reflection* of waves?
b. How are the following wave properties affected by reflection?

frequency   period   wavelength   speed   direction

c. What happens when a parallel beam of waves is incident on an uneven surface?
d. i. What happens when a parallel beam of waves is incident on a smooth plane surface?
   ii. State the relationship between angle of incidence and angle of reflection.
e. Which of our senses relies heavily on reflection of waves? Explain.

Q2
a. What is meant by the term *refraction* of waves?
b. How are the following wave properties affected by refraction?

frequency   period   wavelength   speed   direction

Q3
a. What is meant by the term *diffraction* of waves?
b. How are the following wave properties affected by diffraction?

frequency   period   wavelength   speed   direction

c. What factor(s) affect the degree of diffraction which occurs at an edge?
d. What factor(s) affect the degree of diffraction that occurs when waves pass through a gap or go past an object?
e. Which of our senses relies heavily on diffraction of waves? Explain.

ANSWERS ▶

46 Wave behaviour

1 a Reflection is a characteristic behaviour of waves and occurs when waves bounce off the surface of an object or off the boundary with a different medium.
  b Frequency, period, wavelength and speed are never changed by reflection. The direction of a wave is always changed by reflection.
  c Reflected waves are scattered in many directions.
  d i The waves are reflected as a parallel beam moving in a single direction.
  ii Angle of incidence = angle of reflection.
  e Vision – most objects do not produce their own light – we can see most objects because they reflect light.

2 a Refraction is a characteristic behaviour of waves and occurs when a wave moves from one medium to another.
  b Frequency and period are never changed by refraction. Wavelength and speed are always changed by refraction. Direction is changed when waves are incident at any angle other than perpendicular to the boundary between two media.

3 a Diffraction is a characteristic behaviour of waves and occurs when waves pass through a gap or pass by the edge of an object; some wave energy goes into the shadow region beyond the gap or object.
  b Frequency, period, wavelength and speed are never changed by diffraction. Direction of waves near the edges of an object or a gap is changed by diffraction.
  c At an edge, long wavelength waves diffract more than short wavelength waves.
  d When waves pass through a gap or go past an object, the diffraction is greatest when the width of the gap or object is about the same size as the wavelength of the waves.
  e Hearing: most of the sounds we hear diffract to reach our eardrums.

Exam tip: The answers to the questions above on wave behaviour apply to all kinds of wave so it is worthwhile to make sure you fully understand them.

Higher Physics
Radiation and matter

Interference

Q1 How can we prove that a radiation is a wave?

Q2 What is meant by the term *interference* of waves?

Q3 Waves from two coherent sources overlap in a region of space.
  a What effect do the waves have on each other?
  b What effect do the waves have on the medium?

Q4 a Name the two kinds of interference.
  b Explain how waves from two sources can form bigger amplitude waves.
  c Explain how waves from two sources can form smaller amplitude waves.

Q5 Does interference increase the total energy of waves? Explain your answer.

ANSWERS

47 Interference

1. Set up an experiment to show that an interference pattern can be created using two coherent sources of the radiation.

2. Interference is a characteristic behaviour of waves which occurs when waves from two coherent sources overlap in the same region of space. The overlapping (superposition of waves) results in a pattern of larger and smaller amplitude waves.

3. **a** The waves have no effect on each other – they pass through each other.
   b The waves cause bigger and smaller displacements of the medium.

4. **a** Constructive and destructive.
   b Waves from two sources meet in phase and this results in a larger amplitude wave (crest + crest ⟶ larger crest; trough + trough ⟶ larger trough).
   c Waves from two sources meet out of phase and this results in a smaller amplitude wave or no wave (crest + trough ⟶ smaller or no crest/trough).

5. No. Interference does not change the total energy. There is increased energy at points of constructive interference and decreased energy at points of destructive interference.

Exam **tip:** The answers to the questions above on interference apply to all kinds of wave so it is worthwhile to make sure you fully understand them.

Higher Physics
Radiation and matter

48

Investigating interference

Speed of light in vacuum = $3.00 \times 10^8 \, \text{m s}^{-1}$
Speed of sound in air = $340 \, \text{m s}^{-1}$

Q1
a State the condition for maxima in an interference pattern from two coherent sources.
b State the condition for minima in an interference pattern from two coherent sources.

Q2 Electromagnetic radiation of frequency 1.50×10^{10} Hz passes through a single slit in a metal plate. The radiation then passes through a double slit in a second metal plate.

a What is the function of the double slit?
b What wave behaviour occurs at the double slit and why is this important?
c Calculate the wavelength of the radiation.
d Suggest a suitable width for each slit. Give a reason for your answer.
e What kind of interference occurs at points on the plane perpendicular to the metal plate and midway between the two slits? Justify your answer.
f State three values of path difference for constructive interference.
g State three values of path difference for destructive interference.

Q3 Two loudspeakers are connected to the same output of a signal generator. Starting from a point of maximum of intensity in the middle, a microphone is moved across in front of the loudspeakers. A second maximum of intensity is passed and a third is found at a point which is 2·8 m from one loudspeaker and 4·5 m from the other. Calculate the frequency of the signal generator.

ANSWERS

48 Investigating interference

1 a Maxima occur at points which are a whole number of wavelengths further from one source than the other (*path difference* = $n\lambda$ where *n* is an integer).

b Minima occur at points which are an odd number of half wavelengths further from one source than the other (*path difference* = $(n + ½)\lambda$ where *n* is an integer).

2 a The double slit provides two coherent sources for the radiation.

b Diffraction occurs at the double slit. Diffraction causes waves which pass through each slit to overlap in the region beyond.

c Substituting in $v = f\lambda$
$\Rightarrow \quad 3 \cdot 00 \times 10^8 = 1 \cdot 50 \times 10^{10} \times \lambda$
$\Rightarrow \qquad\qquad \lambda = 0 \cdot 02\,\text{m}\ (2\,\text{cm})$

d 2 cm. Diffraction is greatest when the width of a gap is about the wavelength of the wave.

e Constructive. Points on this plane are the same distance from each slit.

f For constructive interference: 16 cm, 20 cm, 24 cm (any even number).

g For destructive interference: 17 cm, 21 cm, 25 cm (any odd number).

3 Substituting in *path difference* = $n\lambda \Rightarrow (4\cdot5 - 2\cdot8) = 2\lambda$ (*from the wording of the question the path difference must equal two wavelengths*).
$$\Rightarrow \lambda = 0\cdot85\,\text{m}$$
Now substituting in $v = f\lambda \quad \Rightarrow \quad 340 = f \times 0 \cdot 85$
$\qquad\qquad\qquad\qquad\qquad \Rightarrow \quad f = 400\,\text{Hz}$

Exam tip: Make sure you use the correct wave speed; many students have lost marks by using the speed of light in a question about sound or vice versa – do not let it happen to you.

Higher Physics
Radiation and matter

49

Gratings

Q1
a. What is a grating?
b. What wave behaviour occurs as light passes through a grating and why is this important?
c. What wave behaviour occurs when light has passed into the region beyond a grating?

Q2
a. Describe the effect of a grating on a monochromatic light beam.
b. A ray of monochromatic light is incident on a grating. A ray of the same light is incident on a double slit. How is the interference pattern formed by the grating different from the interference pattern formed by the double slit?
c. What name is given to the bright fringe in the centre of the interference pattern?
d. What names are used for the bright fringes on either side of the central fringe?

Q3 Name the quantities in the relationship below and state the SI unit of each.

$$d\sin\theta = n\lambda$$

Q4 A grating with 500 lines per mm is used in an experiment to measure the wavelength of monochromatic light. The following results are obtained:

Average angle of first order max = 14·5°
Average angle of second order max = 30·0°

Use the results to calculate the wavelength of light from the source.

ANSWERS

49 Gratings

1 a A grating is a flat piece of transparent material which has many narrow, parallel slits (lines) very close together and a constant distance apart.
 b Diffraction occurs at the lines. All of the lines act as coherent sources of circular waves. (*Gratings are often called diffraction gratings in physics texts.*)
 c Interference.

2 a An interference pattern of alternate bright and dark areas (fringes) is formed; there is an equal number of bright fringes on each side of a central bright fringe.
 b The interference pattern formed by the grating has fewer, brighter fringes than that formed by the double slit.
 c The bright band in the centre is called the zero order maximum.
 d The bright bands on either side of the centre are called first order maximum, second order maximum, etc. counting out from the zero order.

3 d – spacing of grating lines – metre (m)
θ – angle between centre of a bright fringe and the normal to grating – degree (°)
n – order number of the maximum
λ – wavelength of light – metre (m)

4 For $n = 1$, substituting in $d \sin \theta = n\lambda$

$$\Rightarrow \frac{1 \times 10^{-3}}{500} \times \sin 14 \cdot 5° = \lambda = 500 \, \text{nm}$$

Similarly for $n = 2$, $\frac{1 \times 10^{-3}}{500} \times \sin 30 \cdot 0° = 2\lambda = 1 \times 10^{-6} \, \text{m}$

$$\Rightarrow \lambda = 500 \, \text{nm}$$

***Exam* tip:** In numerical questions on gratings you may be given the separation of adjacent lines or, as in the example above, the number of lines per mm or per cm. Make sure you use the correct value of d in the relationship.

ANSWERS

Higher Physics
Radiation and matter

50

Spectra

Q1
a. Name the colours in the visible spectrum.
b. What natural phenomenon displays the visible spectrum?
c. In this display which colour is closest to and which is furthest from the centre?
d. Which colour has the shortest wavelength and which has the longest wavelength?

Q2
a. State a wavelength for red light.
b. State a wavelength for green light.
c. State a wavelength for blue light.

Q3 A narrow beam of white light is projected through a grating onto a screen.
a. Describe what is observed on the screen.
b. Why is the zero order maximum the colour which is observed?
c. In the first order maxima, which colour is closest to the centre and which colour is furthest from the centre?
d. In the second order maxima which colour is closest to the centre and which colour is furthest from the centre?
e. How is the set of second order maxima different from the set of first order maxima?
f. Which wave property determines the order of the colours?
g. Which wave behaviour(s) is/are responsible for producing the observed effect(s)?

Q4 A narrow beam of white light is projected through a prism onto a screen.
a. Describe what is observed on the screen.
b. Which colour is deviated least and which colour is deviated most?
c. Which wave behaviour(s) is/are responsible for producing the observed effect(s)?

ANSWERS

50 Spectra

1 **a** red, orange, yellow, green, blue, indigo and violet
  b a rainbow
  c nearest the centre – violet; furthest from the centre – red
  d shortest wavelength – violet; longest wavelength – red

2 **a** Any value from 620 to 700 nm (6.2×10^{-7} to 7.0×10^{-7} m).
  b Any value from 490 to 580 nm (4.9×10^{-7} to 5.8×10^{-7} m).
  c Any value from 450 to 490 nm (4.5×10^{-7} to 4.9×10^{-7} m).

3 **a** A symmetrical pattern appears on the screen. The pattern has a narrow white band at the centre with an equal number of visible spectra on either side.
  b The zero order maximum is white as it is made up of the zero order maxima of all of the colours.
  c nearest the centre – violet; furthest from the centre – red
  d nearest the centre – violet; furthest from the centre – red
  e The second order maxima are more spread out than the first order maxima.
  f The colours are arranged in order of wavelength (or frequency).
  g Diffraction (at the grating) and interference are responsible.

4 **a** There is one visible spectrum.
  b deviated least – red; deviated most – violet
  c refraction

***Exam* tip:** If you are asked to compare the white light spectra produced by a grating and a prism, specify *both* similarities *and* differences.

Higher Physics
Radiation and matter

Refraction 1

Q1 Define *refraction of light*.

Q2
  a How is the absolute refractive index of a medium calculated?
  b What implication does this have for the values of absolute refractive indices? Explain your answer.
  c The absolute refractive index of glass A is 1·48 while that of glass B is 1·52. What can you deduce about the speed of light in A compared to the speed of light in B?
  d State a general rule concerning the magnitude of absolute refractive index of a medium and speed of light in that medium.
  e How does the absolute refractive index of a medium compare to the refractive index for light moving from air into the medium?

Q3
  a Sketch a diagram to show the paths followed by two rays of light passing from one medium to another of higher absolute refractive index. One ray should show normal incidence and the other should show incidence at around 40–50°.
  b On your sketch for incidence at around 40–50° label the angle of incidence θ_1 and the angle of refraction θ_2.

Q4 Briefly describe how to measure the refractive index of glass for monochromatic light.

ANSWERS

51 Refraction 1

1. Refraction of light is the change in speed of the light when it moves from one medium to another.

2. **a** Absolute refractive index = $\dfrac{\text{speed of light in a vacuum}}{\text{speed of light in the medium}}$

   b All absolute refractive indices are greater than 1. Light travels faster in a vacuum than in any other medium.

   c Light travels at a greater speed in A than in B.

   d The bigger the absolute refractive index of a medium the slower light travels in the medium (bigger absolute refractive index ⇒ bigger change from c).

   e Absolute refractive index is a fraction bigger; the refractive index for light passing from air to a medium is a good approximation to absolute refractive index.

3. **a** normal incidence **b** incidence at angle θ_1

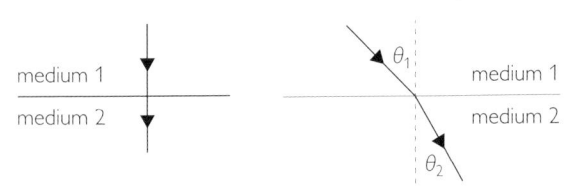

4. Direct a ray of light at a rectangular block of glass placed on a white sheet.

   Mark the position of the incident ray and the entry and exit points of the ray in the glass. Measure angles θ_1 and θ_2.

   Repeat for at least four more angles of incidence.

   For each pair of angles calculate the ratio $\sin\theta_1/\sin\theta_2$. The average value = n.

Exam tip: When a ray bends towards the normal the light gets slower and the wavelength decreases – all the variables get smaller. When a ray bends away from the normal the light gets faster and the wavelength increases – all the variables get bigger.

Higher Physics
Radiation and matter

52

Refraction 2

Q1 Show that absolute refractive index of a medium may be calculated from the wavelength of light in the medium.

Q2 When a narrow beam of white light is projected through a triangular glass prism onto a screen a visible spectrum is observed.

  a Which colour is deviated least? (Remember Topic 50 **Q4b**).
  b For which colour does the prism glass have the smallest absolute refractive index?
  c What do you deduce about the speed in the prism of red light compared to the speed in the prism of all of the other colours of the spectrum?
  d What colour of light is slowest in the prism glass?
  e Where should an infrared detector be placed to detect infrared radiation?
  f What effect does change in medium have on the frequency of light?
  g What do you deduce about colour and degree of refraction?

Q3 Name the quantities in the relationship below and state the SI unit of each.

$$n = \frac{\sin\theta_1}{\sin\theta_2} = \frac{v_1}{v_2} = \frac{\lambda_1}{\lambda_2}$$

ANSWERS

52 Refraction 2

1. Absolute refractive index $n = \dfrac{c}{v_1} = \dfrac{f\lambda}{f\lambda_1} = \dfrac{\lambda}{\lambda_1}$

2. **a** Red light is deviated least.
   b red
   c In the prism glass red light is faster than the light of any other colour.
   d violet
   e The infrared detector should be placed just outside the red end of the spectrum.
   f Change in medium has no effect on frequency.
   g The degree of refraction of light depends on its colour, i.e. its frequency.

3. n – refractive index – this quantity has no unit
   θ_1 – angle of incidence
   θ_2 – angle of refraction – degree (°)
   v_1 – speed of incident light
   v_2 – speed of refracted light – metre per second (m s^{-1})
   λ_1 – wavelength of incident light
   λ_2 – wavelength of refracted light – metre (m)

***Exam* tip:** Like all other waves the frequency of light is determined by the source. Passing light through a prism does not change its frequency; it does change speed and wavelength.

Higher Physics
Radiation and matter

53

Total internal reflection and critical angle

Q1 A girl on a bus at dusk can see an image of the inside of the bus on the windows. When she gets off the bus she can see the inside of the bus through the windows. How is it possible for both of these effects to occur?

Q2 A ray of monochromatic light in a rectangular glass block is incident at 41·80° to the internal surface of the glass. The refractive index of the glass is 1·50.
  a Calculate the angle of refraction.
  b The angle of incidence is now increased by 0·10°. Try to repeat the calculation. Why does the result not make sense?
  c What implication does this have for the refraction of the light?
  d What happens to the ray of light?
  e What happens to the light energy?
  f What name is given to this phenomenon?

Q3 **a** Explain what is meant by the term *critical angle*.
  b Derive the relationship between critical angle and absolute refractive index of a medium.

Q4 **a** Estimate the critical angle for the ray of light in **Q2**.
  b Calculate the critical angle for the monochromatic light and the type of glass in **Q2**.

ANSWERS ▶▶

53 Total internal reflection and critical angle

1 When light in air is incident on a piece of glass some light refracts (i.e. enters the glass) and some reflects (i.e. stays in the air). Light reflected by the window lets the girl see the image of the inside of the bus when she is on the bus and light refracted through the window lets her see inside the bus when she is outside.

2 a Substituting in $\sin\theta_1 = n \sin\theta_2$

$$\Rightarrow \quad \sin\theta_1 = 1.50 \times \sin 41.8° = 0.9998$$
$$\Rightarrow \quad \theta_1 = 88.9°$$

b $\sin\theta_1 = 1.50 \times \sin 41.9° = 1.002$

$\Rightarrow$ result is impossible as maximum sine = 1.

c It is not possible for the ray to refract when incident at 41.9° (or greater) to the normal inside the glass.

d The ray of light is reflected.

e All of the light energy is reflected back inside the glass.

f Total internal reflection.

3 a Critical angle is the angle of incidence when the angle of refraction is 90°.

b $\theta_1 = 90°$, $\theta_2 = \theta_c$ $\quad n = \dfrac{\sin\theta_1}{\sin\theta_2} = \dfrac{\sin 90°}{\sin\theta_c} = \dfrac{1}{\sin\theta_c}$

4 a Any value between 41.8° and 41.9°.

b Substituting in $\sin\theta_c = \dfrac{1}{n}$

$$\Rightarrow \quad \sin\theta_c = \dfrac{1}{1.50}$$
$$\Rightarrow \quad \theta_c = 41.81°.$$

Exam tip: 'Total' because *all* of the energy is reflected. 'Internal' because the energy stays *inside* the glass. 'Reflection' because the light is *reflected*.

Higher Physics
Radiation and matter

Irradiance

Q1
a. Define the term *irradiance*.
b. Name the quantities in the relationship below and state the SI unit of each.
$$I = \frac{P}{A}$$

Q2
a. Name the quantities in the relationship below and state the SI unit of each.
$$I = \frac{k}{d^2}$$
b. Describe the principles of a method to show the relationship between irradiance and distance from a point source.

Q3 A slice of bread 10 cm × 12 cm is placed in an electric toaster. The infrared output of each element of the toaster is 180 W.

a. Estimate the infrared irradiance on each side of the bread.
b. State one assumption you have made in your calculation.

Q4 The irradiance at a distance of 3·5 m from a lamp is 1·5 W m^{-2}. Calculate the irradiance at a distance of 2·5 m.

54 Irradiance

1 a Irradiance at a surface is the power of radiation incident per unit area of the surface.

b I – irradiance – watt per square metre (Wm^{-2})
P – power (of radiation) – watt (W)
A – area – square metre (m^2)

2 a I – irradiance – watt per square metre (Wm^{-2})
k – a constant that depends on the source of radiation – watt (W).
d – distance from source – metre (m)

b Place an irradiance meter and a small source of radiation at the side of a metre stick on top of a black cloth in a darkened room. Note the distance d from the source to the meter and the irradiance reading I on the meter. Adjust the position of the source and again note the measurements. Repeat until at least five pairs of results have been obtained.

Plot a graph of I vs $\frac{1}{d^2}$. A straight line graph passing through the origin is obtained.

3 a $A = 0.10 \times 0.12 = 0.012\,m^2$ $P = 180\,W$
$$I = \frac{P}{A} = \frac{180}{0.012}$$
$$= 1.5 \times 10^4\,Wm^{-2}$$

b This assumes that all of the infrared radiation emitted by an element is incident on the surface of the bread.

4 Substituting in $I = \frac{k}{d^2} \Rightarrow 1.5 = \frac{k}{3.5^2} \Rightarrow k = 1.5 \times 3.5^2\,W$

At $2.5\,m$, $I = \frac{k}{d^2} \Rightarrow I = \frac{1.5 \times 3.5^2}{2.5^2} = 2.9\,Wm^{-2}$

Exam tip: For **Q2b** the experiment is set up on a black cloth to reduce reflection of radiation from the surface of the table or bench. In descriptions of experiments little details which improve accuracy are important.

ANSWERS   * See page 4 re answer layout.

Higher Physics
Radiation and matter

Photons

$h = 6.63 \times 10^{-34}$ Js

Q1
a. Describe a beam of electromagnetic radiation in terms of photons.
b. What determines the energy of an individual photon?

Q2 Name the quantities in the relationship below and state the SI unit of each.

$E = hf$

Q3
a. Calculate the energy of a photon of wavelength 4.75×10^{-7} m.
b. The energy of a photon is 2.50×10^{-18} J. Calculate the frequency of the photon.

Q4
a. Electromagnetic radiation is incident on a metal surface. What happens to the incident photons?
b. Explain irradiance in terms of photons.
c. N photons per second of electromagnetic radiation of frequency f are incident per unit area on a surface. State a relationship for irradiance at the surface due to this radiation.

ANSWERS

55 Photons

1 a A beam of electromagnetic radiation can be regarded as a stream of individual energy bundles called photons, all travelling in the same direction at the same speed.

b Energy of a photon is determined by the frequency of the radiation.

2 E – energy – joule (J)
h – Planck constant – joule second (Js)
f – frequency of electromagnetic radiation – hertz (Hz)

3 a $E = hf = h\dfrac{c}{\lambda} = 6.63 \times 10^{-34} \times \dfrac{3.00 \times 10^8}{4.75 \times 10^{-7}} = 4.19 \times 10^{-19}\,\text{J}$

b Substituting in $E = hf \Rightarrow 2.50 \times 10^{-18} = 6.63 \times 10^{-34} \times f$
$\Rightarrow f = 3.77 \times 10^{15}\,\text{Hz}$

4 a Some photons are reflected and are otherwise unaffected. Other photons are absorbed; each absorbed photon is destroyed and its energy is converted to other form(s) of energy.

b Irradiance at a surface is the sum of the energies of all photons incident on unit area of the surface in one second.

c $I = Nhf$

***Exam* tip:** Photons are an excellent example of the wave–particle duality of the matter in our universe. Photons behave like particles when releasing photoelectrons and behave like waves when forming an interference pattern.

Higher Physics
Radiation and matter

Photoelectric effect

Q1 What is meant by the term *photoelectric effect*?

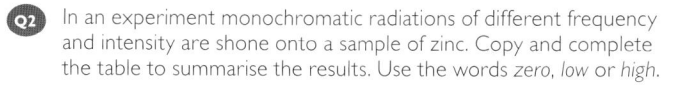

Q2 In an experiment monochromatic radiations of different frequency and intensity are shone onto a sample of zinc. Copy and complete the table to summarise the results. Use the words *zero*, *low* or *high*.

Frequency of radiation	Irradiance at surface	Number of photoelectrons
low	low	
low	high	
high	low	
high	high	

Q3
a What is meant by the term *threshold frequency*?
b What name is given to the minimum photon energy required to eject an electron?
c What happens to the photon energy when a photon with threshold frequency is absorbed by a material?
d What happens to the photon energy when a photon with frequency above the threshold is absorbed by a material?

Q4 Radiation from a monochromatic source of frequency greater than the threshold frequency is incident on a metal surface.

a Do all photoelectrons have the same energy?
b How can you calculate the maximum kinetic energy of a photoelectron?

Q5 For frequencies greater than the threshold value, state the relationship between the photoelectric current produced by monochromatic radiation and the irradiance of the radiation at the surface.

ANSWERS

56 Photoelectric effect

1 The photoelectric effect is a naturally occurring phenomenon in which electromagnetic radiation incident on a metal surface causes electrons to be ejected from the surface.

2

Frequency of radiation	Irradiance at surface	Number of photoelectrons
low	low	zero
low	high	zero
high	low	low
high	high	high

3 a Threshold frequency is the minimum frequency of radiation able to cause the emission of photoelectrons from a surface.
  b Work function.
  c All of the photon energy is used to release one electron from the surface.
  d Some photon energy is used to release one electron from the surface and the remainder is converted to kinetic energy of that electron.

4 a No, photoelectrons have a range of kinetic energies.
  b Maximum electron kinetic energy = (photon energy − work function)

5 For frequencies greater than the threshold value, the photoelectric current produced by monochromatic radiation is directly proportional to the irradiance of the radiation at the surface.

***Exam* tip:** The condition in the answer to **Q5** is important – for frequencies less than the threshold value there is no photoelectric current. Make sure you include the condition if you want full marks.

Higher Physics
Radiation and matter

Energy levels in atoms and line spectra

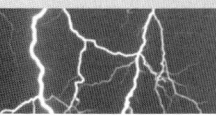

Q1
- **a** Can electrons in atoms have any value of energy?
- **b** Name the lowest electron energy level in an atom.
- **c** Name the energy level at which an electron is released from an atom.
- **d** What is meant by the term *excited state*?

Q2
- **a** Describe one way in which an electron may be raised to an excited state.
- **b** What condition must be met for an electron to be raised from ground level to an excited state?
- **c** How long do electrons stay in excited states?

Q3 White light is passed through a sample of an element and projected through a triangular prism onto a screen.
- **a** Describe what happens to photons while the light is passing through the sample.
- **b** What effect does this have on the white light spectrum?

Q4
- **a** What happens to the energy of an electron when it falls from an excited state to a lower energy level in the atom?
- **b** Do electrons in excited states always fall straight to the lowest energy level?

Q5 Atoms of sodium are excited electrically in a discharge tube. Light from the discharge tube is passed through a grating and the spectrum is observed.
- **a** What happens to the atoms of sodium?
- **b** Describe the spectrum observed.

ANSWERS

57 Energy levels in atoms and line spectra

1 a No – only certain energy values are possible.
  b ground state
  c ionisation level
  d An excited state is an electron energy level between ground level and ionisation level. In an excited state an electron has more energy than in the ground state, but not enough energy for ionisation.

2 a An electron may be raised to an excited state electrically, by heating or by absorption of a photon.
  b The energy absorbed by the electron must exactly equal the difference in energy between the ground state and the excited state.
  c Normally a very short time; the time is random and an electron may fall to a lower level at any time.

3 a Photons with energies exactly equal to the difference between two electron energy levels are absorbed.
  b The spectrum has a number of dark lines corresponding to the frequencies of the absorbed photons.

4 a The energy is used to create a photon of electromagnetic radiation which is emitted by the atom.
  b No, an electron in an excited state may fall to any lower energy level.

5 a Electrons in the atoms are raised to excited states and these then fall to lower levels, thus emitting photons.
  b The spectrum consists of a series of bright lines corresponding to the frequencies of the emitted photons.

***Exam* tip:** If you are asked to draw a diagram of the energy levels in a hydrogen atom, get the spacing right – the biggest gap is between the ground state and the level above – each subsequent gap is smaller than the previous gap.

Higher Physics
Radiation and matter

58

Lasers

Q1 What are the characteristics of spontaneous emission of a photon when an electron in an atom falls from one energy level to another?

Q2 What are the characteristics of stimulated emission?

Q3 What does the acronym *laser* stand for?

Q4 Describe the functions of the mirrors in a laser.

Q5 Explain why even low-power lasers are dangerous.

ANSWERS

58 Lasers

1 Spontaneous emission is random. There is no external cause and it may happen at any time. Emitted photons may have any permitted energy and may travel in any direction.

2 When a photon of energy equal to an atomic energy level difference is incident on an atom in an excited state, the photon may stimulate an electron to fall to a lower level and emit a second photon. The incident photon and emitted photon have the same frequency, are in phase and travel in the same direction.

3 *Light Amplification by the Stimulated Emission of Radiation*

4 A laser has a mirror at either end. The function of the mirrors is to keep the photons inside the laser material so that more stimulated emissions can occur. One mirror is completely reflecting. The other is partially reflecting and allows a small proportion of light energy to pass through.

5 A low-power laser beam can be dangerous because the edges of a laser beam are very close to parallel. The irradiance of the beam is nearly constant. It does not decrease in the same way as the irradiance of light from a bulb. (*Remember a beam of laser light having a power even as low as 0·1 mW might cause eye damage – so be safe!*)

***Exam* tip:** A laser gains energy by stimulated emission and loses energy by absorption; lasers are designed to that the energy gained is greater than the energy lost – this is what makes a constant beam possible.

Higher Physics
Radiation and matter

Semiconductors

Q1
a. Describe the electrical properties of conductors.
b. Describe the electrical properties of insulators.
c. Describe the electrical properties of semiconductors.

Q2
a. What is *doping*?
b. What effect does this have on the electrical properties of a semiconductor?

Q3
a. How many valence electrons does an atom of silicon have?
b. How is n-type silicon produced?
c. What are the main charge carriers in n-type semiconductor materials?
d. How is p-type silicon produced?
e. What are the main charge carriers in p-type semiconductor materials?

Q4
a. Is n-type semiconductor material negatively charged? Explain your answer.
b. Is p-type semiconductor material positively charged? Explain your answer.

ANSWERS

59 Semiconductors

1 a Conductors have a low resistance and allow current to pass easily.

  b Insulators have a very high resistance and do not let current pass unless a very high voltage is applied.

  c Semiconductors have intermediate resistance values.

2 a Doping is a process for adding a tiny number of impurity atoms to a semiconductor.

  b Doping decreases the resistance of the semiconductor material.

3 a four

  b Pure silicon is doped with a tiny number of atoms of an element with five valence electrons (e.g. nitrogen, phosphorus). Four electrons of each doping atom form covalent bonds with silicon electrons. One electron is 'extra'.

  c electrons

  d Pure silicon is doped with a tiny number of atoms of an element with three valence electrons (e.g. boron, aluminium). The electrons of each doping atom form covalent bonds with silicon electrons. The fourth chemical bond cannot be completed and this leaves a 'hole' in the crystal structure.

  e holes

4 a No, n-type semiconductor material is electrically neutral. Though there are extra electrons within the crystal structure there are exactly the same number of 'extra' protons in the doping atoms.

  b No, p-type semiconductor material is electrically neutral. Though there are missing electrons (holes) within the crystal structure there are exactly the same number of missing protons in the doping atoms.

***Exam* tip:** Holes do not actually move but it can be useful to think of holes as positive charge carriers moving in the opposite direction to electrons.

Higher Physics
Radiation and matter

p–n junction diode

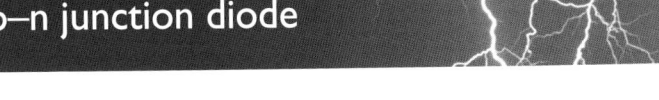

Q1
a. How is a p–n junction diode made?
b. What happens at the junction between the p-type and n-type semiconductor?
c. Here is the symbol for a diode:

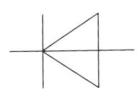

Explain how to interpret the symbol.

Q2
a. How is a diode made forward-biased?
b. What happens inside the semiconductor materials when the diode is forward-biased?
c. What effect does this have regarding charge carriers in the diode?
d. Does a forward-biased diode conduct?

Q3
a. How is a diode made reverse-biased?
b. What effect does this have regarding charge carriers in the diode?
c. Does a reverse-biased diode conduct?

60 p–n junction diode

1 a A p–n junction diode is made by joining of a piece of p-type semiconductor to a piece of n-type semiconductor.

  b 'Extra' electrons in the n-type cross the junction and fill holes in the p-type. As a result, the region near the junction has no charge carriers. The n-type material near the junction becomes positively charged and the p-type material near the junction becomes negatively charged.

  c The 'arrow' shape in the symbol points from p-type to n-type.

2 a A diode is forward-biased when the n-type is connected to the negative terminal of a supply and the p-type is connected to the positive terminal of the supply.

  b In the n-type, electrons from the negative terminal replace 'extra' electrons that have crossed to the p-type to fill holes. In the p-type, electrons attracted towards the positive terminal are pulled out of filled holes.

  c There are charge carriers throughout the diode. Electrons move towards the positive terminal and holes move towards the negative terminal.

  d Yes, a forward-biased diode conducts.

3 a A diode is reverse-biased when the n-type is connected to the positive terminal of a supply and the p-type is connected to the negative terminal of the supply.

  b The size of the region with no charge carriers is increased.

  c No, a reverse-biased diode does not conduct.

***Exam* tip:** In a diode the region near the boundary between p-type and n-type semiconductor material is known as the depletion layer. As the p type is negative and the n-type is positive there is a p.d. (about 0·6 V for silicon) across this layer even when the diode is not connected in a circuit. In a forward-biased diode the e.m.f. of the supply acts in the opposite direction to the depletion layer p.d. and must overcome it for conduction to occur.

Higher Physics
Radiation and matter

Applications of diodes

Q1
a. What is a light emitting diode (LED)?
b. How is an LED connected in a circuit?
c. Describe how light is produced in an LED.
d. Give one application of an LED.

Q2
a. What is a photodiode?
b. What is the source of energy in a photodiode circuit?
c. In what mode does the diode operate when it acts as a photodiode?

Q3
a. A photodiode is reverse-biased in a circuit containing a supply and a resistor. In what mode is the photodiode operating?
b. Describe what happens in the circuit when light is incident on the p–n junction.
c. The irradiance of the light incident on the diode is doubled. What effect does this have?
d. The voltage of the supply is increased. What effect does this have?
e. Give one application of a photodiode used in this way.

ANSWERS

61 Applications of diodes

1 a An LED is a diode designed to emit a particular colour (frequency) of light.
  b forward-biased
  c In the junction region positive and negative charge carriers recombine to emit quanta of radiation.
  d Indicator light in an electronic circuit.

2 a A photodiode is a diode which produces positive and negative charges when light is incident on the p–n junction.
  b Light energy incident on the p–n junction is the source of energy in the circuit.
  c photovoltaic mode

3 a photoconductive mode
  b Light incident on the p–n junction causes a small (leakage) current.
  c The leakage current doubles.
  d This has no significant effect on the leakage current.
  e Bar-code scanner, fast switch in telecommunications equipment.

Exam **tip:** For a photodiode in photoconductive mode, current is fairly independent of the reverse-biasing voltage at voltages lower than the breakdown voltage (the voltage at which the diode is forced to conduct).

Higher Physics
Radiation and matter

62

n-channel enhancement MOSFET

Q1 What does the acronym *MOSFET* stand for?

Q2 Explain the electrical *off* state of an n-channel enhancement MOSFET.

Q3 Explain the electrical *on* state of an n-channel enhancement MOSFET.

Q4 Why does the name include the term *n-channel*?

Q5 Draw the circuit symbol for an n-channel enhancement MOSFET. Label the source, the gate and the drain.

Q6 State one application for an n-channel enhancement MOSFET.

ANSWERS

62 n-channel enhancement MOSFET

1 Metal Oxide Semiconductor Field Effect Transistor.

2 An n-channel enhancement MOSFET is connected in a circuit so that the drain is made more positive than the source.

 As the back electrode is connected to the source, the junction between the substrate and the source is unbiased.

 The junction between the drain and the substrate is reverse-biased.

 This reverse bias and the high resistance of the substrate prevent the flow of charge.

3 To switch on the MOSFET a positive voltage of over 2 volts is applied to the gate.

 This sets up an electric field between the gate and the back electrode.

 Electrons in the p-type substrate gather in a layer beneath the gate and form a channel.

 The channel enables charge to flow from the source to the drain.

4 The layer of negatively charged electrons in the p-type substrate form a channel – hence the name n-channel.

5

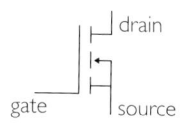

[circuit symbol diagram with labels: gate, drain, source]

6 Amplifier.

***Exam* tip:** For your exam you are expected to be able to 'describe' the structure of an n-channel enhancement MOSFET – the easiest way to do this may be to draw a labelled diagram.

Higher Physics
Radiation and matter

The atom and nucleus

Q1 A beam of alpha particles is fired in a vacuum at a thin sheet of gold.
  a State two key observations from this experiment.
  b What do these observations suggest regarding the structure of gold atoms?
  c What did Rutherford propose regarding the structure of atoms?

Q2 Explain the meanings of the letters in the symbol below.

$$^A_Z X$$

Q3
  a State three types of naturally occurring radioactive decay.
  b Which part(s) of an atom are affected by radioactive decay?
  c In what order were the three types of radioactive decay discovered?
  d What effect(s) do these radiations have when they pass through matter?
  e Which of the radiations can be stopped completely?
  f Can the decay of an individual nucleus be predicted?

Q4
  a What effect(s) does alpha decay have on a nucleus?
  b What effect(s) does beta decay have on a nucleus?
  c What effect(s) does gamma decay have on a nucleus?

ANSWERS

63 The atom and nucleus

1 a Most of the alpha particles pass straight through the sheet undeflected. A small number of alpha particles bounce back in the direction from which they came.
  b The first observation suggests that most of a gold atom is empty space. The second suggests that gold atoms contain very small, massive and positively charged objects.
  c Rutherford proposed an atomic model in which:
  most of the mass of an atom is concentrated in a tiny massive core – the nucleus; the diameter of the nucleus is much smaller than the diameter of the atom; electrons orbit the nucleus.

2 A is the mass number = number of protons + number of neutrons.
Z is the atomic number = number of protons.
X is the chemical symbol of the element.

3 a alpha, beta and gamma.
  b the nucleus
  c Alpha was discovered first, followed by beta and then gamma.
  d They all cause ionisation of atoms which are near to their paths.
  e alpha and beta only
  f No, the decay of an individual nucleus is random and may occur at any time.

4 a The energy level of the nucleus is decreased, its mass number is decreased by 4 and its atomic number is decreased by 2.
  b The energy level of the nucleus is decreased, its mass number remains the same and its atomic number is increased by 1.
  c The energy level of the nucleus is decreased, its mass number and atomic number remain the same.

***Exam* tip:** Alpha, beta and gamma are the first three letters of the Greek alphabet. When these radiations were discovered they were called the equivalent of A, B and C because physicists did not know what they were. The greatest ionisation is caused by alpha, which is why it was discovered first. Less ionisation is caused by beta and even less by gamma.

Higher Physics
Radiation and matter

Nuclear reactions

Q1
a. Write out alpha decay in symbolic form.
b. Write out beta decay in symbolic form.
c. Write out gamma decay in symbolic form.

Q2
a. What is meant by the term *fission*?
b. What are the characteristics of naturally occurring fission?
c. What are the characteristics of fission reactions in nuclear reactors?
d. What happens to a large nucleus when it absorbs an additional neutron?
e. How does the total mass of the products of a fission reaction compare to the total mass of the reactants?
f. How do you account for this?

Q3
a. What is meant by the term *fusion*?
b. How does the total mass of the products of a fusion reaction compare to the total mass of the reactants?
c. How does this reaction generate heat energy?

Q4
a. Name the quantities in the relationship below and state the SI unit of each.
$E = mc^2$
b. Calculate the energy released when 1 kg of mass is converted to energy.

ANSWERS

64 Nuclear reactions

1 **a** $_Z^A X \longrightarrow \,_{Z-2}^{A-4} W + \,_2^4 He$
(X $\longrightarrow$ W, the new nucleus is a different chemical element)

b $_Z^A X \longrightarrow \,_{Z+1}^{A} Y + \,_{-1}^{0} e$
(X $\longrightarrow$ Y, the new nucleus is a different chemical element)

c $_Z^A X \longrightarrow \,_Z^A X + \gamma$ photon
(X $\longrightarrow$ X, the new nucleus is the same chemical element)

2 **a** In fission a nucleus of large mass number splits into two nuclei of smaller mass number, usually with the release of a small number of energetic neutrons.

b Naturally occurring fission is usually spontaneous and random.

c Fission reactions in reactors are induced by bombarding large nuclei with neutrons. A chain reaction results.

d A large nucleus becomes unstable and a fission reaction occurs.

e Whether spontaneous or induced, the products of a fission reaction have smaller mass than the reactants.

f The missing mass is converted to kinetic energy of the products.

3 **a** In fusion, two nuclei of small mass number combine to form a nucleus of larger mass number.

b The products of a fusion reaction have smaller mass than the reactants.

c The mass which is lost is converted to kinetic energy of the products.

4 **a** E – energy released by nuclear reaction – joule (J)
m – mass converted to energy – kilogram (kg)
c – speed of light in a vacuum – metre per second (m s^{-1})

b Substituting in $E = mc^2 \Rightarrow E = 1 \times (3 \times 10^8)^2 = 9 \times 10^{16}$ J

***Exam* tip:** In your exam you should be able to identify the processes occurring in nuclear reactions written in symbolic form – this includes fission and fusion reactions, as well as alpha, beta and gamma decay.

Higher Physics
Radiation and matter

Dosimetry

Q1
a. Define the *activity* of a radioactive source.
b. Name the SI unit of activity of a radioactive source.
c. Express the SI unit of activity in decays per second.

Q2
a. Define the quantity *absorbed dose*.
b. Name the SI unit of absorbed dose.
c. Express the SI unit of absorbed dose in terms of other SI units.

Q3
a. What factors affect the risk of biological harm to tissue?
b. How is this risk accounted for in calculations on exposure to radiation?

Q4
a. Define the quantity *equivalent dose*.
b. Name the SI unit of equivalent dose.
c. Define the quantity *equivalent dose rate*.

Q5 What is *effective dose*?

ANSWERS

65 Dosimetry

1 a The activity of a radioactive source is the average number of nuclei decaying per unit time.
   b becquerel (Bq)
   c 1 Bq = 1 decay per second (1 decay s^{-1}).

2 a When tissue is exposed to radiation, the absorbed dose is the energy absorbed per unit mass of the tissue.
   b gray (Gy)
   c 1 Gy = 1 joule per kilogram (J kg^{-1}).

3 a The risk of biological harm to tissue depends on the absorbed dose, the kind of radiation, and the body organs or tissue exposed.
   b A radiation weighting factor w_R is given to each radiation; this factor is a measure of the biological effect of the radiation.

4 a Equivalent dose is the product of absorbed dose and radiation weighting factor.
   b sievert (Sv)
   c Equivalent dose rate is the equivalent dose per unit time.

5 Effective dose is the sum of the equivalent doses for all body tissues exposed to radiation and takes into account the susceptibilities to harm of each type of body tissue. Effective dose indicates the overall risk to health from exposure to ionising radiation.

Exam tip: For your Higher exam you are expected to be able to perform calculations using all of the dosimetry relationships which are found towards the end of the list of relationships for Higher Physics – make sure you practise these skills before the exam.

Higher Physics
Radiation and matter

Background radiation

Q1 What is *background radiation*?

Q2
- **a** Name and describe a radiation which comes from outer space and which contributes to background radiation.
- **b** What features of Earth give us protection from this radiation?
- **c** Do these features give us complete protection?

Q3
- **a** What type(s) of radiation from sources on Earth contribute(s) to background radiation?
- **b** Is the level of radiation from sources on Earth constant across the planet?

Q4
- **a** Name a living source of background radiation.
- **b** Explain why this source is radioactive.
- **c** Name a radioactive isotope in living things.

Q5
- **a** Name a source of background radiation which is dangerous in buildings.
- **b** Why is this source dangerous?

Q6 State the value of the average annual effective dose of background radiation in the UK?

ANSWERS

66 Background radiation

1 Background radiation is radiation to which we are exposed because of our place in space on planet Earth.

2 a Cosmic rays: these are high-energy particles, mostly from the Sun, that bombard the Earth.
  b The atmosphere and the Earth's magnetic field protect us from cosmic rays.
  c No, some cosmic rays reach the Earth's surface.

3 a Alpha, beta and gamma radiations from radioactive elements in our environment.
  b No, the level of this radiation depends on the number and type of radioactive elements in the rocks, soil, water and air in a location.

4 a Humans, animals and plants are all sources of background radiation.
  b People and animals eat, drink and breathe radioactive elements in our surroundings. Plants absorb radioactive elements through water and from the air.
  c carbon-14

5 a radon gas
  b Radon gas is dense and sometimes accumulates in the basement of buildings.

6 The average effective dose of background radiation in the UK = 2 mSv per year.

Higher Physics
Radiation and matter

Radiation and safety

Q1 a What limit has the government set for the annual effective dose for exposure of the general public to radiation?
b What limit has the government set for people who work in industries that use radioactive materials?

Q2 a What is meant by the term *shielding*?
b What factors affect the effectiveness of shielding?

Q3 Sketch a graph to show how the irradiance of a beam of gamma radiation varies with the thickness of an absorber.

Q4 a What is meant by the term *half-value thickness*?
b Describe the principles of a method for measuring the half-value thickness of an absorber.

Q5 Describe one other method of reducing equivalent dose rate.

ANSWERS

67 Radiation and safety

1. **a** 1 mSv per year
   b 20 mSv per year
2. **a** Shielding reduces exposure by placing absorbing material in the path of radiation.
   b The effectiveness of shielding depends on the type of material and the thickness of this material.
3.

   Graph: irradiance (y-axis) vs thickness of absorber (x-axis), showing a decreasing exponential curve.

4. **a** The half-value thickness of an absorber for a particular radiation is the thickness that absorbs half of the radiation.
   b Measure the level of background radiation using a detector and counter.
   Place a source a fixed distance in front of the detector. Measure the count rate.
   Place a measured thickness of absorbing material between the source and detector. Measure the count rate again.
   Repeat the procedure with additional measured thicknesses of absorber until sufficient measurements have been made.
   Subtract the background count rate from each measured value of count rate.
   Plot a graph of corrected count rate *vs* thickness of absorber.
   Analyse the graph to find the average thickness of absorber which halves the count rate.
5. Equivalent dose rate is also reduced by increasing the distance from sources of radiation.

***Exam* tip:** Make sure you understand the precautions to be taken when handling radioactive sources – this is not just important for your exam, it may also be important for your health.

Higher Physics
Course skills

Numbers and units

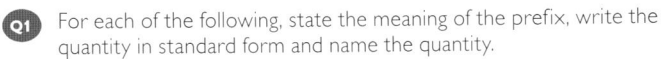

Q1 For each of the following, state the meaning of the prefix, write the quantity in standard form and name the quantity.

- **a** 3·9 GJ
- **b** 25 pF
- **c** 400 kΩ
- **d** 100 mm
- **e** 510 nm
- **f** 48 µGy
- **g** 5·0 MW

Q2 State the abbreviation of the SI unit for each of the following quantities.

- **a** depth
- **b** gravitational field strength
- **c** tension
- **d** impulse
- **e** voltage gain
- **f** Planck constant
- **g** effective dose
- **h** refractive index
- **i** half-life

Q3 What is the difference, if any, between ms^{-1} and m s^{-1}?

Q4
- **a** How many figures should you include in your final answer to a numerical calculation?
- **b** How many figures should you keep in intermediate values used in calculating your final answer?

ANSWERS

68 Numbers and units

1 a G – giga : $3.9\,\text{GJ} = 3.9 \times 10^9\,\text{J}$: energy
 b p – pico : $25\,\text{pF} = 25 \times 10^{-12}\,\text{F} = 2.5 \times 10^{-11}\,\text{F}$: capacitance
 c k – kilo : $400\,\text{k}\Omega = 400 \times 10^3\,\Omega = 4.00 \times 10^5\,\Omega$: resistance
 d m – milli : $100\,\text{mm} = 100 \times 10^{-3}\,\text{m} = 1.00 \times 10^{-1}\,\text{m}$; length or distance or displacement
 e n – nano : $510\,\text{nm} = 510 \times 10^{-9}\,\text{m} = 5.10 \times 10^{-7}\,\text{m}$: wavelength
 f μ – micro : $48\,\mu\text{Gy} = 48 \times 10^{-6}\,\text{Gy} = 4.8 \times 10^{-5}\,\text{Gy}$: absorbed dose
 g M – mega : $5.0\,\text{MW} = 5.0 \times 10^6\,\text{W}$: power

2 a m
 b $\text{N}\,\text{kg}^{-1}$
 c N
 d Ns
 e no unit
 f Js
 g Sv
 h no unit
 i s

3 $\text{ms}^{-1} \equiv \dfrac{1}{\text{ms}}$ $\text{m s}^{-1} \equiv \dfrac{m}{s}$ *(the symbol ≡ means 'is equivalent to')*

4 a The number of figures (i.e. the number of significant figures) in a final answer should be the same as the minimum number of significant figures given in the data used to calculate the answer.
 b The number of figures in an intermediate answer should be one more than the number of figures in the final answer.

***Exam* tip**: Get into the habit of using correct SI units every time you tackle a numerical problem. This habit will save you time and will help you get units correct.

Higher Physics
Course skills

Uncertainties

Q1
a. What is meant by the term *uncertainty*?
b. What is the main aim of experimental design regarding uncertainties?
c. What factors can contribute to inaccuracy in individual measurements?
d. What is the most serious effect that uncertainties can have on experimental results?

Q2 In an electrical experiment the p.d. across a resistor was measured as 4·0 V. The experimenter estimated that the uncertainty of this measurement was 0·2 V.

a. Write down this result as a measurement and an absolute uncertainty.
b. Write down this result as a measurement and a percentage uncertainty.
c. State the relationship between absolute uncertainty and percentage uncertainty.

Q3
a. Five different quantities are measured in an experiment. The absolute uncertainty in each measurement is estimated. How is the measurement with the greatest effect on the accuracy of the experiment identified?
b. Copy and complete the following statement:
   'In an _____ the _____ percentage _____ in any _____ measurement is often a good _____ of the _____ uncertainty in the _____ numerical result.'

ANSWERS

69 Uncertainties

1 a Uncertainty is a measure of the degree of accuracy of an experimental measurement.

b The main aim of experimental design regarding uncertainties is to minimise uncertainty and so maximise the accuracy of the final experimental results.

c Degree of accuracy of a physical measurement can be affected by:
- the maximum accuracy of available measuring equipment
- the conditions under which the measurement takes place
- the care and skill of person(s) carrying out the experiment
- the stability of the quantity being measured
- the stability of environmental conditions which may affect experimental variables.

d The most serious effect of uncertainty is to invalidate the results of an experiment (for example a physical measurement is meaningless if the uncertainty in the measurement is greater than the measurement itself).

2 a p.d. across resistor = 4.0 ± 0.2 V

b p.d. across resistor = $(4.0 \pm 5\%)$ V

c percentage uncertainty = $\dfrac{\text{absolute uncertainty}}{\text{measurement}} \times 100\%$

3 a Calculate the percentage uncertainty in each measurement. The measurement with the highest percentage uncertainty has the greatest effect on the accuracy of the experiment.

b 'In an *experiment* the *largest* percentage *uncertainty* in any one measurement is often a good *estimate* of the *percentage* uncertainty in the *final* numerical result.'

***Exam* tip:** Questions on the evaluation of experiments often centre on reduction of uncertainty in experimental measurements.

ANSWERS

Higher Physics
Course skills

Random and systematic uncertainties

Q1
- a. What is meant by the term *random uncertainty*?
- b. How can random uncertainties be eliminated from experimental measurement?
- c. Describe a method for reducing the effect of random uncertainty.
- d. What is the best estimate of the true value of a quantity with random uncertainty in its measurement?

Q2
- a. Describe how to calculate the mean of a number of measurements.
- b. Describe how to estimate the random uncertainty in the mean of a number of measurements.

Q3
- a. State the scale-reading uncertainty of an analogue meter.
- b. State the scale-reading uncertainty of a digital scale.

Q4
- a. What is meant by the term *systematic uncertainty*?
- b. How can effects of systematic uncertainties be identified?
- c. What effect does systematic uncertainty have on the mean value of a measured physical quantity?
- d. Mary and John are setting up circuits to measure resistance using an ammeter and a voltmeter. The circuits are shown below.

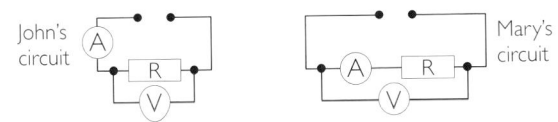

John's circuit / Mary's circuit

Has either student eliminated systematic uncertainty? Explain your answer.

Q5 Students often complain that 'experiments do not work'. Is there another explanation?

ANSWERS

70 Random and systematic uncertainties

1 a Random uncertainty has unpredictable effects and so measurements may be too large or too small.
  b Random uncertainties *cannot* be eliminated from an experiment.
  c Repeating measurements and calculating a mean reduces the effect of random uncertainty and leads to greater accuracy.
  d The mean is the best estimate of the true value of the quantity.

2 a Add the measured values and divide by the number of measurements.
  b Provided at least five measurements have been taken, the difference between the maximum and minimum measured values divided by the number of measurements is a good estimate of the random uncertainty in the mean.

3 a For analogue scales the scale-reading uncertainty is usually ± half of the smallest scale division.
  b For digital scales the scale-reading uncertainty is ±1 in the last digit.

4 a Systematic uncertainty has a consistent effect on the measurement of a quantity; all measured values may be either too big or too small.
  b Effects of systematic uncertainties can be identified by plotting a graph.
  c Where there is a systematic effect the mean is offset from the true value.
  d Neither student has eliminated systematic uncertainty. In John's circuit the ammeter measures the total current in the resistor and in the voltmeter so the current reading is consistently high. In Mary's circuit the voltmeter measures the p.d. across the ammeter and the resistor so the voltmeter reading is consistently high.

5 Experiments always work – the physical properties of materials under investigation always follow the laws of the universe. When an experiment produces an 'incorrect' result it is highly likely that the actual uncertainties are so high that the results are meaningless.

Exam tip: Scale-reading uncertainty is the minimum uncertainty in the quantity being measured. The actual uncertainty is often much greater.